DISTRACTED

THE EROSION OF ATTENTION
AND THE COMING DARK AGE

DISTRACTED

MAGGIE JACKSON
Foreword by BILL McKIBBEN

Prometheus Books
59 John Glenn Drive
Amherst, New York 14228–2119

Published 2009 by Prometheus Books

Inquiries should be addressed to
Prometheus Books
59 John Glenn Drive
Amherst, New York 14228–2119
VOICE: 716–691–0133, ext. 210
FAX: 716–691–0137
WWW.PROMETHEUSBOOKS.COM

12 11 10 09 08   5 4 3 2 1

Cataloging-in-Publication Data on file at the Library of Congress

Jackson, Maggie, 1960–
    Distracted : the erosion of attention and the coming Dark Age / Maggie Jackson.
       p. cm.
    Includes bibliographical references and index.
    ISBN 978–1–59102–623–5 (hardcover)
    ISBN 978–1–59102–748–5 (paperback)
    1. United States—Civilization—1970– 2. United States—Social conditions—
1980– 3. United States—Social life and customs—1971– 4. Distraction
(Psychology)—United States. 5. Attention—Social aspects—United States.
6. Technology—Social aspects—United States. 7. United States—Civilization—21st
century—Forecasting. 8. Social change—United States. 9. Social psychology—United
States. I. Title.

E169.12.J26 2008
306.0973—dc22

2008016540

Printed in the United States of America on acid-free paper

For Emma and Anna

CONTENTS

FOREWORD

As I settled down at my desk to write this brief foreword, a light on the computer blinked to indicate that a new e-mail had arrived. This left me with a quandary that by now must afflict most Americans most days of their lives: continue with the train of thought that I'd begun to follow or see who was hailing me and for what purpose. The quandary was easily resolved—I of course clicked on my inbox, finding a message from someone asking if she could call me right now to discuss a piece of climate change legislation making its way through the Senate. I said yes, the phone rang, a half hour disappeared, and now I am back.

I can't add up with any precision the costs and benefits of that small exchange, repeated constantly. On the one hand, I was able to link up with someone far more easily than I would have even a few years ago. On the other hand, I was unable to keep thinking a thought —any thought—straight through.

Distraction has always been a human condition. Sages have always been quick to point out how even a few minutes of meditation prove the jumpy nature of our consciousness—our monkey minds. But now every force conspires to magnify that inattentiveness: technology has made distraction ubiquitous. We're almost always within reach of something to fill our brains—how often do you sit in the car without turning on the radio? How often do you enter a hotel room without turning on the TV or, now, looking for the Ethernet cable? Places that once offered some respite—the coffee shop, the waiting room—are now among the most connected.

Since this is the water we swim in, it's hard to notice it unless some artificial condition interferes. Some artificial condition like . . . nature. The occasional blackout is a nuisance, but also a gift. A long solo backpacking trip is a revelation—for two or three days your internal CNN continues to chatter away, but at a certain point it starts to run out of opinions, plans, screeds, and begins to fall silent a little. It's weird, unsettling.

This book, remarkably impressive both for its wealth of detail and the clarity of its synthesis, forces our attention on that inattention. And in so doing, it asks us implicitly the uncomfortable question about what our lives are for. Are they measured in busyness, the accomplishment of many and random tasks? Or do they require some kind of more artful arc to be whole? Writing powerfully and subversively, Maggie Jackson raises issues that go straight to the core of what it means to be human in the early twenty-first century, questions that we need to think about clearly, slowly, deeply.

The inbox is flashing again, clamoring for my attention. Loving novelty, I head in its direction; craving depth, I do so with a tinge of regret.

—Bill McKibben

INTRODUCTION

I am alone inside a glass booth, enveloped in a swirling storm of dialogue snippets, data shards, and clashing storylines. This is a play, for an audience of one. Within the booth is a tiny cybercafe, with a table, two chairs, a fake daisy in a vase, and a Mac computer. Sit down, press a button, and the play begins—disembodied, prerecorded, unshared. Over a soundtrack of clinking dishes and laughing customers, I hear two hushed voices, talking as if from another table. A man and a woman are meeting secretly, worried that their spouses are having a cyber-affair. The Mac comes to life and unfurls an instant message chat: the virtual whispers of the cyber-lovers. I listen and read, as the two couples simultaneously, online and offline, debate whether cyber-love is real, whether harm has been done. "We haven't cheated, if that's what you're thinking," says one. "Then why do I feel guilty?" his virtual lover retorts. The parallel scripts wash over me, clashing and overlapping, competing for my gaze and for my ear.

I listen twice and then try the next booth, and the next. There are six in all, a half dozen playlets, each no more than twelve minutes long, told by and through computers. These are the Technology Plays—six faceless, mechanized dramas of passion, mistrust, and dehumanization, unfolding at the touch of a button.[1] This smorgasbord of experimental theater is on display in the soaring atrium of a college library in upstate New York. Just before the plays are dismantled, I have driven three hours to see them. Here on this dreary campus on a raw November day, I am looking for clues to understanding our own increasingly multilayered, mutable, and virtual world. Are these six mini-plays absurd slices of science fiction or reflections of our own lives of snippets and sound bites? Are they meaningless theatrical peep shows or dramatizations of the real shadows creeping over our lives?

Enter Booth 2, Richard Dresser's *Greetings from the Home Office*. Now I am sitting at an office desk, cast as the new hire in a corporate office that's roiled by a possible accounting scandal. A telephone rings. A frantic woman colleague is calling to say that the boss is covering up his misdoings. The smooth-talking boss telephones next to say that the whistleblower should be ignored; she is his bitter ex-lover. "I'm saying this for your own good," the boss purrs. Rapid-fire calls are punctuated by the interjections of a meddling secretary on an intercom. Ultimately, I'm forced to decide whom to trust, based on incomplete information and faceless relations. It's the age-old game of office politics done blindfolded, execution-style. In another play, *Chip*, a seemingly banal ATM transaction goes wrong when the machine rejects the identity chip allegedly implanted in my finger. The machine and a bank representative on the telephone shout contradictory commands as I hear machine-gun-firing security guards close in. The experience is at once comic and frightening. When the machine orders me not to turn around, I can't help but give a terrified peek over my shoulder. The Technology Plays are equal parts video game, amusement park ride, and high-tech storytelling. Oh-so briefly, you are sent hurtling into fantastical worlds where you are asked to fix what you can't see, decide what you don't understand, relate to those you cannot fully trust. The narratives are fragmented. The experience is disorienting. This is not how we live. Or is it?

I close the door of the last booth and walk back to my car. I've talked to no one all afternoon except a faceless woman whom I called on my cell phone about a technological glitch. (A man wordlessly comes to fix it.) The next day is Thanksgiving. The campus is silent and nearly deserted, with only a few backpack-toting students milling about. In twenty-four hours, most will be gathering with family, as will I. But going forward, I can't shake the lingering unease left by these fragments of plays. They manage to distill in brief much of how we live: tossed and turned by info-floods, pummeled by clashing streams of rapid-fire imagery, floating in limitless cyber-worlds, all while trust and privacy and face-to-face moments slip from our grasp. Twelve-minute plays for a one-minute world. Push the button and the next morsel of "real life" unfolds. Guess again, a soupçon of information will suffice. Do you dare peek over your shoulder? Can you clearly see the way ahead?

We can tap into 50 million Web sites, 1.8 million books in print, 75 million blogs, and other snowstorms of information,[2] but we increasingly seek knowledge in Google searches and Yahoo! headlines that we gulp on the run while juggling other tasks. We can contact millions of people across the globe, yet we increasingly connect with even our most intimate friends and family via instant messaging, virtual visits, and fleeting meetings that are rescheduled a half dozen times, then punctuated when they do occur by pings and beeps and multitasking. Amid the glittering promise of our new technologies and the wondrous potential of our scientific gains, we are nurturing a culture of social diffusion, intellectual fragmentation, sensory detachment. In this new world, something is amiss. And that something is attention.

The premise of this book is simple. The way we live is eroding our capacity for deep, sustained, perceptive attention—the building block of intimacy, wisdom, and cultural progress. Moreover, this disintegration may come at great cost to ourselves and to society. Put most simply, attention defines us and is the bedrock of society. Attention "is the taking possession by the mind, in clear and vivid form, of one out of what seem several simultaneously possible objects or trains of thought," wrote psychologist and philosopher William James in 1890. "It implies withdrawal from some things in order to deal effectively with others, and is a condition which has a real opposite in the confused, dazed, scatter-brained state which in French is called *distraction*, and *Zerstreutheit* in German."3 James came tantalizingly close to understanding at least one aspect of this mysterious phenomenon whose inner workings eluded philosophers, artists, historians, and scientists for centuries. But today, we know much more about attention, and all that we are learning only serves to underscore its irrefutable importance in life. Attention is an organ system, akin to our respiratory or circulation systems, according to cognitive neuroscientist Michael Posner. Attention, as James astutely understood, is the brain's conductor, leading the orchestration of our minds. And its various networks are key not only to higher forms of thinking but to our morality and even our very happiness.

Yet increasingly, we are shaped by distraction. James described a clear and vivid possessing of the mind, an ordering, and a withdrawal. We easily recognize that these states of mind are becoming less and less a given in our lives. The seduction of alternative virtual universes, the addictive allure of multitasking people and things, our near-religious allegiance to a constant state of motion: these are markers of a land of distraction, in which our old conceptions of space, time, and place have been shattered. This is why we are less and less able to see, hear, and comprehend what's relevant and permanent, why so many of us feel that we can barely keep our heads above water, and our days are marked by perpetual loose ends. What's more, the waning of our powers of attention is occurring at such a rate and in so many areas of life, that the erosion is reaching critical mass. We are on the verge of losing our capacity as a society for deep, sustained focus. In short, we are slipping toward a new dark age.

A dark age? Certainly, the notion seems so far removed from our techno-miraculous era that the mere mention of the subject appears irresponsibly alarmist. Amid our material riches, abundant information, and creative leaps, how can we be headed toward a time of decline? How can I speak of cultural slippages when we are able to decipher the genome, map the ocean floor, and plumb the depths of the brain to begin understanding something as abstract and intangible as attention? But take a closer look. The parallels between dark ages past and our times are clear and multiplying, and the erosion of attention is the key to understanding why we are on the cusp of a time of widespread cultural and social losses.

Consider first that a dark age is not a one-dimensional time of unending disintegration. Rather, it is a distinct turning point in history, a period of flux that often produces great technological and other gains yet ultimately results in a declining civilization and a desertlike spell of collective forgetting. The medieval era was a fertile time technologically, marked by the invention of eyeglasses, glazed windows, fireplaces, windmills, and stirrups, along with the compass, mechanical clock, and rudder.[4] Yet by the sixth century, all of continental Europe's great libraries not only had disappeared but the *memory* of them was lost to an emerging feudal society, notes Thomas Cahill.[5] After the fall of the mighty Mycenaean Empire in the tenth century BC, the Greeks improved upon the great rowing ships of the time and upon military tactics such as the phalanx. For the first time, the olive was cultivated as a food. Yet overall living standards slipped dramatically, and advances in carving, building, the arts, and farming slipped away as Greece entered a five-hundred-year-long dark age. Beads of amber came to be made of bone. Painted murals and elaborately carved gems gave way to clay modeling. The dead were buried in the crumbling ruins of beautiful buildings. Even writing declined in the time between the waning of the script called "Linear B," and the adoption of an alphabet.[6] A dark age can dazzle if you fail to see the whole. But inexorably, such a time leads to "a culture's dead end."

Chillingly, many of these shifts parallel the current trajectory of our civilization. I'm not arguing that we'll soon be living in wooden huts and burying our dead in ruined landmarks, yet we, too, are in the midst of a time of innovations, flux, and impending decline. Civilization is defined by a "sense of permanence," notes historian Kenneth Clark. "Civilized man, or so it seems to me, must feel that he belongs somewhere in space and time."[7] Consciously doing all we can to free ourselves from the last boundaries of space and time, we are ultimately trading our cultural and societal anchors for an age of glorious freedom, technological innovation—and darkness. As our attentional skills are squandered, we are plunging into a culture of mistrust, skimming, and a dehumanizing merging between man and machine. As we cultivate lives of distraction, we are losing our capacity to create and preserve wisdom and slipping toward a time of ignorance that is paradoxically born amid an abundance of information and connectivity. Our tools transport us, our inventions are impressive, but our sense of perspective and shared vision shrivel. This is why Umberto Eco has called Western society "neomedieval," comparing our own anxious, postliterate, and highly mobile age with the first post-Roman millennium in Europe. The late Jane Jacobs, who wrote the classic *Death and Life of Great American Cities* a generation ago, warned of an impending time of darkness brought on by fatal cracks in various pillars of American society, such as ethics and science. Critic Harold Bloom calls this era a "Dark Age of the Screens."[8]

Although a time of decline may take generations or even centuries to unfold, the first evidence of its shadowy embrace may be seen all around us, if we can rise above the muddle of our days and see clearly. With this in mind, I want to lure you, the reader, into pausing and reflecting on our daily lives and our era. I do not propose to, nor could I, write *The Decline and Fall of the American Empire*, nor could I tackle the enormous political or economic implications of such changes. Nor is this a how-to, although hopefully this book will have a positive impact on a reader's life. Rather, my scope is *how we live and what that means* for our individual and collective futures.

Consider the issue of attention deficit disorder. This is not a book about whether millions of distractible children and adults belong on Ritalin and other stimulants. Yet alongside the voracious debates about the disorder festers a discomfiting realization: this condition, or at least some portion of it, may be exacerbated by the way we are living. With a mixture of self-mockery, perverse pride in our busyness, and real uneasiness, we accept that ADD has spilled far past the confines of the medical world. Educators fret about "vaccinating" children against ADD through better teaching and parenting. A growing roster of defenders venerate the upsides of being attention-deficient. Students pop Adderall and other "wakefulness" pills to focus better on tests. "How do you know you have ADD or a severe case of modern life?" asks Edward Hallowell, a psychiatrist who warns of an epidemic of attention deficit trait. "Everyone these days is super-busy and multitasking and keeping track of more data points than ever. The actual condition is just that—taken to a more extreme level."[9] ADD and ADT are both causes of worry and badges of honor, but above all, they signal disturbing slippages in an asset that we cannot afford to lose.

Not long ago, an old school friend called me. We hadn't talked in years. After hearing about my book, he wanted to get together. E-mails sallied back and forth, like bumper cars bouncing off one another, with no solid hits. After repeated cancellations and postponements, we set a date for lunch. Arriving at his office, I waited while he wound up a meeting, took a phone call, dashed around. (Luckily, I had brought a book.) It was long past noon when we hurried to a coffee shop, where he mentioned that he'd eaten already. I gulped a sandwich, he half-heartedly sipped a soda, then raced back to work. It was all very friendly, yet I stood on the street corner for a few minutes, feeling unsettled. In school, I was studious and he was goofy. But he was never this, well, *diluted*. I thought we were going to catch up. Instead, I got the face-to-face equivalent of an MTV montage. Several times, he uneasily joked about his inability to focus.

We all share the joke. Nearly a third of workers feel they often do not have time to reflect on or process the work they do. More than half typically have to juggle too many tasks simultaneously and/or are so often interrupted that they find it difficult to get work done.[10] One yearlong study found that workers not only switch tasks every three minutes during their workday but that nearly half the time they interrupt themselves.[11] People are so dazed that they have almost no time to reflect on the world around them, much less their futures. Not too long ago, I stood on a New York street corner, watching a businessman who was so busy shouting into a cell phone and hailing a taxi that for some minutes he didn't notice that his colleague already had found a cab and was ready to roll. Speaking at a lunch for forty lawyers in Philadelphia, I talked about our mobile lives, technology's impact on the home, flexible work—and the erosion of focus. They pounced. That was the lightning rod issue: divided, diluted attention. Two hours passed, without a BlackBerry in sight. Finally, the leader of the group tentatively asked, "Is it just me, or is it true that we don't seem to go deeply into anything anymore?" Around the table, heads nodded. "Is all of this really progress?" she wondered.

Is all this progress? We have reason to worry. Kids are the inveterate multitaskers, the technologically fluent new breed that is better suited for the lightning-paced, many-threaded digital world, right? After all, they are bathed in an average of nearly six hours a day of nonprint media content, and a quarter of that time they are using more than one screen, dial, or channel. Nearly a third of fourteen- to twenty-one-year-olds juggle five to eight media while doing homework.[12] Yet for all their tech fluency, kids show less patience, skepticism, tenacity, and skill than adults in navigating the Web, all while overestimating their prowess, studies show.[13] Meanwhile, US fifteen-year-olds rank twenty-fourth out of twenty-nine developed countries on an Organization for Economic Cooperation and Development (OECD) test of problem-solving skills related to analytic reasoning—the sort of skills demanded in today's workforce. Nearly 60 percent of fifteen-year-olds in our country score at or below the most *basic* level of problem solving, which involves using single sources of well-defined information to solve challenges such as plotting a route on a map.[14] Government and other studies show that many US high school students can't synthesize or assess information, express complex thoughts, or analyze arguments. In other words, they often lack the critical thinking skills that are the bedrock of an informed citizenry and the foundation of scientific and other advancements.[15] Is it just a coincidence that they lack the skills nurtured by the brain's executive attention network, the seat of our highest-level powers of focus?

While undoubtedly the reasons for this state of affairs are myriad, what's certain is that we can't be a nation of reflective, analytic problem solvers while cultivating a culture of distraction. I am not alone in wondering how often our children will experience the hard-fought pleasures of plunging deeply into a thought, a conversation, a state of being. Will focusing become a lost art, quaintly exhibited alongside blacksmithing at the historic village? ("Look, honey, that man in twentieth-century costume is doing just one thing!") An executive at a top accounting firm, who was researching the future workforce, confided to me his deep concerns that young workers are less and less able to concentrate, think deeply, or mine a vein of inquiry. Knowledge work can't be done in sound bites, he warned.

And yet again, perhaps it is silly to worry. What if we're all slowly, and with lots of stumbles, learning to adapt to a new world, where multitasking, split-second decision making, information surfing, bullet pointing, and virtual romancing is the new norm? What if our technologies are ushering us into an even more glorious future? Just hang on for the ride! Perhaps the smart folks are creating a new definition of smarts. Those clinging to the power of the word, the mysterious miracles of the human mind and spirit, and a reverence for the sensuality of physical togetherness will be soon left behind!

It's true that our tools are shaping us and teaching us, as they have since man first picked up a stick, sword, or pen. And today's technologies are making us smarter . . . in some ways. We don't yet know whether video games teach us problem-solving, pattern recognition, or how to construct order from chaos, as Steven Johnson and others argue.[16] Early evidence shows that many computer games teach the kind of iconic and spatial skills useful for playing more computer games, and not much more. While some carefully crafted computer-based exercises can and are being powerfully used to educate, many games on the market are simply "training wheels" for improving *computer* literacy, some researchers conclude. Video games also can boost your ability to pay attention to multiple stimuli in your field of vision, a skill useful for driving but not analytical thinking.[17] Limited doses of educational television produce some language, math, and reading gains in young children, but there's no evidence that, as Johnson argues, the growing numbers of sitcom characters or "complexity" of plot lines in adult fare offer us anything more than a diet of loose threads.[18] If crafted well and used wisely, our high-tech tools have the potential to make us smarter, but too often so far they seem to be doing so in narrow ways.

"It's like, instead of writing the whole book, you're just writing the back cover," explained Brendan, age ten.[19] One sultry June day in library class, Brendan was giving me a primer on the difference between a written report—"writing, research, more research, more writing"—and his PowerPoint presentation on piranha fish—"a brief summary." Although Brendan was hardly a boy of few words in person, his slides took bullet points to new heights of brevity. "Habitat. Amazon River. Warm Water. Murky Water. Always Have Food. Lots of Predators," one read. He ticked off the benefits of PowerPoint. It's shorter, animated, fun, and easier, even though you don't learn as much from it. His classmates and other students at a suburban elementary school outside Hartford, Connecticut, agreed with him. They liked PowerPoint's short sentences, bouncing words, the fact that their parents are amazed by the results, and above all, its bright colors. No one wants to read long sentences or too much information, they assured me. Not surprisingly, the information in their presentations was mostly cut and pasted from Web sites. The results? "You get a smattering of information," said the school librarian, whose job is to teach tech skills, not content. "You get a snippet of this and that." When I asked several children about specific words in their slides or why they included a particular and sometimes nonsensical piece of information, most shrugged. "I don't know," they said.

Channeling our thoughts and ideas into outlines and bullet points— is this a new definition of smart? With four hundred million copies in circulation, PowerPoint is the world's most popular presentation tool. Some books get taken off class lists if they don't give good Power-Point.[20] Many corporate employees, bureaucrats, and military officers hardly dare to make a presentation without it. Proponents, including Microsoft (naturally), argue that in a world of information overload, we need brevity and distillation. PowerPoint's shortcomings stem from people's inexperience with this new tool, they say. But opponents, led by information theorist Edward Tufte, are blistering in their criticism. They say the software shapes thoughts and data into alluringly professional yet narrow and simplistic formats, fostering uninformative and often misleading presentations that discourage creativity or argument. Partly as a result of Tufte's work, NASA blamed its reliance on Power-Point for a pattern of obfuscation and miscommunication that helped doom the *Challenger* space shuttle.[21] "When presentation becomes its own powerful idea, we diminish our appreciation of complexity," concluded MIT professor Sherry Turkle after studying PowerPoint's use in classrooms.[22] PowerPoint "tells you how to think," quips rock musician David Byrne in "Envisioning Emotional Epistemological Information," a satirical PowerPoint artwork set to his own compositions. "This can be an enjoyable experience."[23]

PowerPoint doesn't make us dumb, and a judicious use of such slideware likely can help us navigate a world of information overload. I'm not angling for a return to some sort of pastoral, unmechanized Eden in order to halt the erosion of attention. We cannot blame technology for society's ills. Nor can we fall into the opposite and increasingly commonplace trap of blindly trusting that our new tools will automatically usher us into a glorious new age. The tools we are wholeheartedly embracing today are inherently powerful, and we ignore that truth at our peril. You can use a stick for digging potatoes or stabbing your neighbor, so how you use a stick is important, but equally important is the fact that a stick is not a wheel. It's crucial that we better understand how our new high-tech tools, from video games to PowerPoint, may be affecting us. Moreover, our tools reflect the values of our time, so it's no coincidence that PowerPoint is a tool of choice in a world of snippets and sound bites. This is the messy soup that makes up our relation to technology, and explains why technology plays a starring but ultimately subordinate role in this book. Technology is a key to understanding our world, but it is not the full story. Instead, we must ask: how do we want to define progress? We are adapting to a new world, but in doing so are we redefining "smart" to mostly mean twitch speed, multitasking, and bullet points? Are we similarly redefining intimacy and trust? Are we so enamored of our alluring surface gains that we are failing to stem or even notice the deeper, human costs of these "advances"? We may be getting expert at pushing buttons and wiring our thoughts with bells and whistles and tracking two virtual enemies across a screen but losing the skills needed to thrive in an increasingly complex world: deep learning, reasoning, and problem solving. We may be getting better at juggling three people simultaneously over a screen or wire but forgetting that what we need more than ever in a time of growing mistrust and seemingly expendable relations is honesty, unhurried presence, and care.

I think we're beginning to see a time of darkness when, amid a plethora of high-tech connectivity, one-quarter of Americans say they have no close confidante, more than double the number twenty years ago. It's a darkening time when we think togetherness means keeping one eye, hand, or ear on our gadgets, ever ready to tune into another channel of life, when we begin to turn to robots to tend to the sick and the old, when doctors listen to patients on average for just eighteen seconds before interrupting, and when two-thirds of children under six live in homes that keep the television on half or more of the time, an environment linked to attention deficiencies. We should worry when we have the world at our fingertips but half of Americans age eighteen to twenty-four can't find New York state on a map and more than 60 percent can't similarly locate Iraq.[24] We should be concerned when we sense that short-term thinking in the workplace eclipses intellectual pattern making, and when we're staking our cultural memory largely on digital data that is disappearing at astounding rates. We should worry when attention slips through our fingers.

For nothing is more central to creating a flourishing society built upon learning, contentment, caring, morality, reflection, and spirit than attention. As humans, we are formed to pay attention. Without it, we simply would not survive. Just as our respiration or circulatory systems are made up of multiple parts, so attention encompasses three "networks" related to different aspects of awareness, focus, and planning.[25] In a nutshell, "alerting" makes us sensitive to incoming stimuli, while the "orienting" network helps us select information from among the millions of sensations we receive from the world, voluntarily or in reaction to our surroundings. A baby's first job is to hone these skills, which are akin to "awareness" and "focus," respectively. In a class of its own, however, is the executive network, the system of attention responsible for complex cognitive and emotional operations and especially for resolving conflicts between different areas of the brain. (We fire up four separate areas of the brain just to solve a simple word recognition problem, such as coming up with a use for the word *hammer*.)[26] All three networks are crucial and often work together, and without strong skills of attention, we are buffeted by the world and hindered in our capacity to grow and even to enjoy life. People who focus well report feeling less fear, frustration, and sadness day to day, partly because they can literally deploy their attention away from negatives in life. In contrast, attentional problems are one of the main impediments to attaining "flow," the deep sense of contentment that people find when they are stretching themselves to meet a challenge.[27] For example, schizophrenics tend to suffer from *anhedonia*, or literally "lack of pleasure," because they usually are unable to sift stimuli. "Things are coming in too fast," said one patient. "I am attending to everything at once and as a result I do not really attend to anything."[28] Without a symphonic conductor, the music of the brain disintegrates into cacophonic noise.

Attention also tames our inner beast. Primates that receive training in attention become less aggressive.[29] One of attention's highest forms is "effortful control," which involves the ability to shift focus deliberately, engage in planning, and regulate one's impulses. Six- and seven-year-olds who score high in tests of this skill are more empathetic, better able to feel guilt and shame, and less aggressive. Moreover, effortful control is integral to developing a conscience, researchers are discovering.[30] In order to put back the stolen cookie, you must attend to your uneasy feelings, the action itself, and the abstract moral principles—then make the right response. All in all, attention is key to both our free will as individuals and our ability to subordinate ourselves to a greater good. The *Oxford English Dictionary* defines attention as "the act, fact or state of attending or giving heed; earnest direction of the mind," and secondarily as "practical consideration, observant care, notice." The word is rooted in the Latin words *ad* and *tendere*, meaning to "stretch toward," implying effort and intention.[31] Even the phrase "attention span" literally means a kind of bridge, a reaching across in order to widen one's horizons. Attention is not always effortful, but it carries us toward our highest goals, however we define them. A culture that settles for numb distraction cannot shape its future.

This morning, I sat in the library trying to harness my thoughts, but like runaway horses, they would not be reined in. We missed the first birthday of a baby whose parents are two of our closest friends, and I stewed about that for a while. Someone I interviewed for my newspaper column was peeved that an editor hadn't confirmed this evening's photo shoot, and a flurry of e-mails ensued. A man outside my study room made a long, illicit phone call and gestured threateningly when I asked him to stop. The old water pipes hissed in a new, odd way. I have never thought of myself as easily distracted, although I always have had an Olympic capacity for daydreaming, failing to complete my sentences, and slips of the tongue. (Go put on your pajamas, I'll tell my daughter at noon, when I really mean bathing suit.) A friend generously diagnoses me as having "too many words" in my head. But perhaps the ultimate weak spot in my own capacity for focus is that I'm observant and "sensitive"—a label I grew to hate as a child. I notice much of the ceaseless swirl of social intonations, bodily signals, and facial expressions around me. Even in the relative quiet of the library, the world tends to come rushing in, jumbling and splashing about inside me, a restless sea.

Yet isn't that essentially the starting point of attention? Attention is a process of taking in, sorting and shaping, planning, and decision making—a mental and emotional forming and kneading of the bread of life, or, if you prefer, an inner mountain climb. The first two forms of attention—alertness and orienting—allow us to sense and respond to our environment, while the third and highest network of executive attention is needed to make ultimate sense of our world. Our ability to attend is partly genetic, yet also dependent upon a nurturing environment and how willing we are to reach for the highest levels of this skill, just as a naturally gifted athlete who lacks the opportunity, encouragement, and sheer will to practice can never master a sport. Today, our virtual, split-screen, and nomadic era is eroding opportunities for deep focus, awareness, and reflection. As a result, we face a real risk of societal decline. But there is much room for hope, for attention can be trained, taught, and shaped, a discovery that offers the key to living fully in a tech-saturated world. We need not waste our potential for reaching the heights of attention. We don't have to settle for lives mired in detachment, fragmentation, diffusion. A *renaissance of attention* is within our grasp.

I didn't set out to write about attention. I was curious why so many Americans are deeply dissatisfied with life, feeling stressed, and often powerless to shape their futures in a country of such abundant resources. At first, I sought clues in the past, assuming that lessons from the first high-tech era—the heyday of the telegraph, cinema, and railway—could teach us how to better manage our own shifting experiences of space and time. Instead, I discovered that our gadgets are bringing to a climax the changes seeded in these first revolutions. Is this a historical turning point, a "hinge of history," in Thomas Cahill's words? In researching this next question, I discovered stunning similarities between past dark ages and our own era. At the same time, I began studying the astonishing discoveries made just in our own generation about the nature and workings of attention. As I explored these seemingly unrelated threads, I realized that they formed a tapestry: the story of what happens when we allow our powers of attention to slip through our fingers. Realizing this loss is intriguing. Considering the consequences is alarming.

When a civilization wearies, notes Cahill, a confidence based on order and balance is lost, and without such anchors, people begin to return to an era of shadows and fear.[32] Godlike amid our five hundred television channels and three hundred choices of cereal, are we failing to note the creeping arrival of a time of impermanence and uncertainty? Mesmerized by streams of media-borne eye candy and numbed by our faith in technology to cure all ills, are we blind to the realization that our society's progress, in important ways, is a shimmering mirage? Consumed by the vast time and energy simply required to survive the ever-increasing complexity of our systems of living, are we missing the slow extinction of our capacity to think and feel and bond deeply? We just might be too busy, wired, split-focused, and *distracted* to notice a return to an era of shadows and fear.

On the August day in 410 when the Goths brutally sacked Rome, the emperor was at his country house on the Adriatic, attending to his beloved flock of prize poultry.[33] Informed by a servant that Rome had perished, the emperor, Honorius, was stunned. "Rome perished?" he said. "It is not an hour since she was feeding out of my hand." The chamberlain clarified himself. He'd been talking about the city, not the imperial bird of that name. Apocryphal as this story may be, the point is apt. For in the years leading into a dark age, societies often exhibit an inability to perceive or act upon a looming threat, such as a declining resource. Twilight cultures begin to show a preference for veneer and form, not depth and content; a stubborn blindness to the consequences of actions, from the leadership on down. In other words, an epidemic erosion of attention is a sure sign of an impending dark age.

Welcome to the land of distraction.

Part I
LENGTHENING SHADOWS
Exploring Our Landscape of Distraction

Chapter One

WIRED LOVE CA. 1880

Tracing the Roots of an Attention-Deficient Culture

"I hope sometime we may clasp hands bodily as we do now spiritually, on the wire—for we do, don't we," said C.

"Certainly, here is mine spiritually!" responded Nattie without the least hesitation, as she thought of the miles of safe distance between.

—Wired Love: A Romance of Dots and Dashes[1]

The quaint words, like tinges of a foreign accent, date this prose. This fictional scene is set in an era of horse-drawn carriages, hoop skirts, and chaperoned outings. Two young telegraph operators begin quarreling, then chatting, over the wire and eventually fall in love after a long flirtatious exchange of messages and innuendo, all heard by other operators up and down the same telegraph line. When they finally meet, they are so tongue-tied and distracted by jealousies and other real-world complications that they only get engaged after setting up a private telegraph line between rooms in a boardinghouse so they can tap out their amorous feelings. Their story unfolds in *Wired Love: A Romance of Dots and Dashes*, an 1880 novel that ends with a Morse code proposal from "C" to "N."

Wired Love is a long-forgotten literary footnote, a peek into a time far removed from the whirling, speeding Technology Plays, when dusty rail journeys and "stereopticon" slide-viewers were considered state of the art. But this minor portrait is worth resurrecting if we are to understand how we arrived at an age of splintered attention, on the cusp of a dark time. For although we breezily assume that cell phones, e-mail, and instant messages inspired our split-screen, split-second life, we must remember that virtual love didn't start with the Internet, and jets and cars didn't entirely make us the nomads we are today. Dizzying film and video montages trace their ancestry to wisps of Victorian prose and poetry detailing lives that now strike us as gritty and plodding. The era captured in *Wired Love* signaled a panoply of heady, new powers for humankind that seemed to break our earthly fetters, shift our world—and set off changes that ushered us toward a land of distraction. Simultaneity, or the seeming ability to exist in two places at once, rewrote the nature of human communication and set the stage for our fevered attempts to juggle multiple, mediated domains. Spiritualism, with its earnest efforts to tap into new frontiers of communication, foreshadowed our notions of cyberspace as a realm of scientific, spiritual, and social exploration. Last, the era's earnest experimentation with mediated experience and the control of perception has helped inure us to a world of fragmented, diffused, and manipulated attention. To understand our age, we must first look back. "The world is stretched & folds out & squeezes shut like an accordion tortured by some sadistic hand," wrote Blaise Cendrars, an early twentieth-century French poet who tried to capture the revolutionary changes of his day. "Through the rents in the sky the locomotives hurl themselves in fury. . . . I'm afraid I've no idea how to see it through to the end."[2]

To appreciate the emotional impact of the telegraph and then the telephone, remember that for at least one thousand years before such devices were invented, a message could be transmitted at most just one hundred miles by horseman in a day, notes Tom Standage.[3] Simultaneous communication over a distance could not be imagined.

Nothing in the natural world anticipated such communications, note James Katz and Mark Aakhus in *Perpetual Contact*, an exploration of the impact of the cell phone. Even the all-powerful mythic Greek gods had to rely on messengers, they write.[4] News of George Washington's death in Virginia took a week to reach New York, while nearly 70 percent of the country knew that John F. Kennedy had been assassinated within a half hour of the Dallas shooting.[5] More than two thousand people died in 1815 at the Battle of Orleans—*two weeks after* a peace treaty had been signed in London between the combatants.[6] In 1825, Samuel Morse missed the funeral of his wife, Lucretia, in New Haven because at the time he was in Washington, DC—a four-day journey from Connecticut. But by the 1870s, about twenty years after Morse had created the first public telegraph, six hundred and fifty thousand miles of wire and thirty thousand miles of submarine cable had been laid, and a message could be sent from London to Bombay and back in as little as four minutes, notes Standage in his history of the telegraph, *The Victorian Internet*.[7]

The effect of such innovations on people's sense of time and space was electrifying, as evidenced by the fin de siècle and early twentieth-century love affair with "simultaneity" of all kinds. This was a time of stunning, discomfiting advances not just in communications but in travel and entertainment. The railroad, steamship, telephone, and other "homeless technologies" made it seem possible to be everywhere at once, noted German cultural historian Karl Lamprecht in 1912.[8] Speed became an often lamented norm. Electricity lit up the night. And the phonograph and cinema played still more havoc with sensory experience, seemingly stretching, bending, and freezing once-stable conceptions of time and space. "Present-day life, more fragmented and faster-moving than preceding periods, was bound to accept as its means of expression, an art of dynamic 'divisionism,'" wrote cubist painter Fernand Léger in a 1913 essay exploring the roots of the brief, influential art movement that Pablo Picasso and Georges Braque founded.[9] Just as the cubists tossed away established conventions of beauty and perspective to paint fractured abstract collages inspired by the mind's eye, so Virginia Woolf and other writers shelved conventional rules of narrative to try to capture the rapid-transit simultaneity of both the material and psychological worlds. Simultaneity's literary master-work, of course, was James Joyce's *Ulysses*, a cinematic pastiche of moments in a multitude of lives touched by one man's brief journey through Dublin. In the book, a mid-June day becomes "a vast expanded present," observes historian Stephen Kern.[10]

Little more than a decade before the publication of *Ulysses*, Cendrars published an intriguing and prescient little work that plays with a marriage of simultaneity in both idea and form. Claimed by the poet to be the first "simultaneous book," the 1913 *La Prose du Transsibérien et de la petite Jehanne de France* depicts his 1904 journey from Moscow to Harbin, spiced with musings of his mistress. Illustrated with geometric splashes of color down one side and with a map of the journey across the top, the poem was printed on a six-foot-long sheet of paper that was intended to be unfolded and seen all at once. Although thoroughly low-tech by our standards, the piece nevertheless still gives readers a real sense of speed, movement, and the collapse of distance. At that time, most books rarely experimented with chronology and invariably were illustrated with stodgy picture plates distinct from the text. But Sonia Delaunay's illustrations swirl alongside the poem, with swatches of rose, mint, midnight blue, and tangerine blending into one another, seemingly impossible to confine. The poem is sparsely punctuated, and nearly every word sports a different hue or font size. A reader is pushed and pulled from ancient to modern times and places, and back again.

Que mon couer
tour à tour, brûlait comme le temple d'ephèse
ou comme la Place Rouge de Moscou quand le soleil se couche.

(That my heart
one moment was aflame like the temple at Ephesus, & the next
was like the sun in its sinking on Moscow's Red Square.)

While Cendrars revels in the dizzying changes of the day, *Wired Love*'s author Ella Cheever Thayer depicts the awkwardness and uncertainties that complicated the seductive new relations of this age. Thayer, a hotel telegraph operator in Boston, points out at the novel's outset that her protagonist, nineteen-year-old Nattie Rogers, lives in two worlds: her drab rented room in a stuffy dowager's apartment and the telegraph office where she could escape "on a sort of electric wings, to distant cities and towns," and where "although alone all day, she did not lack social intercourse."[11] In the novel's opening scene, Nattie multitasks between these worlds furiously, juggling the barrage of code thrown at her by "C" with the ignorant questions of a curious spectator and the demands of a pushy, would-be customer. Throughout the book, Nattie tends to scowl at customers standing in her office, preferring the allure of the wire and especially her witty invisible friend to the demands of the physical world in front of her. Yet, like the cyber-lovers in the Technology Play, Nattie wrestles constantly with the question of whether wired love is real. She wants to believe it is, but can't quite get over her doubts. Flirting with Clem ("C"), she hopes uneasily that the physical distance between them excuses their improper familiarities. Even after they get to know each other by telegraph, they distrust the reality of their relationship so completely that when they meet, they are essentially strangers and must go back to square one in the courtship game. They seek refuge in a private telegraph wire, but wind up promising to end their wired romance in favor of "real" love, "more potent than electricity."

Virtual love affairs were often played out in actual telegraph offices, where women, agile at both machine and code, made up a good portion of the clubby corps of operators of the day. By the 1870s, women comprised a third of the staff at Western Union's main office in New York.[12] Telegraph companies tried to break up wired romances and friendships by "snatching," or transferring operators from station to station. But between telegrams, the operators in different stations found ways to keep playing virtual checkers, telling stories, and flirting. "Romances of the Telegraph," an 1891 article in the weekly

Western Electrician told the story of a male operator who struck up a virtual friendship with an operator whom he assumed to be a man. After the "fellow" backed out of a planned vacation together, the two nevertheless arranged a meeting that ultimately led to courtship and marriage. In another instance, a remote operator was refused leave to travel for his marriage, so he arranged to hold the ceremony over the wire. A minister in California officiated to the operator and his bride in Arizona. "At the proper moment, the solemn 'I do' came back over the wires," the weekly reported. "Congratulatory messages were sent the bride and groom from all stations."[13]

Virtual love affairs and online games? Simultaneous books and instantaneous travel? Cubist painters, simultaneous poets, and telegraph operators were undoubtedly the front line of industrialized society's plunge into a culture of simultaneity and split-screen attention. They were the first to confront questions that we now wrestle with daily. How do we navigate seen and unseen worlds at once and handle relationships that unfold via new codes of disembodiment and mystery? How is humanity changed when, as Kern notes, distance collapses and the present becomes "an extended interval of time that could, indeed must, include events around the world"?[14] We still wrestle with the protocols and parameters of the virtual, while increasingly embracing it as a *real* new space for human creation and connection. At the same time, simultaneity is morphing into something richer and more layered. Just as the shocking glimpse of a Victorian ankle has melted into today's blasé public exhibitions of thigh, belly, and décolletage, we've moved from the once novel experience of hopscotching distance to share a moment, to the jaded expectation of juggling multiple temporal and spatial dimensions. Teens describe instant messaging their peers as less enjoyable than face-to-face visits or phone calls, yet they say that they prefer IM because it allows them to multitask each other. "Personally, I like talking to a lot of people at a time," a Pittsburgh-area teen told researchers. "It kind of keeps you busy. It's kind of boring just talking to one person cause then like . . . you can't talk to anyone else."[15] Now we weave in and out of a vast array of relationships, dancing across multiple spaces of connection, seemingly freed from the limits of body and earth. Attention becomes ethereal in a world of multiplicity. No longer do boundaries matter.

Along with simultaneity, cyberspace's precursor was the realm of spiritualism, the nineteenth-century movement to communicate with the dead. Flourishing just a few years after the invention of the telegraph, spiritualism's practices ranged from fraudulent circus-like séances to earnest experiments and writings on the subject by great thinkers and scientists of the day, including Thomas Edison, Rudyard Kipling, Mark Twain, and Sir Arthur Conan Doyle. Thomas Edison devoted an entire chapter in his memoir to communicating with the dead.[16] Twain was a member of Britain's Society for Psychical Research and in an 1884 article on "thought transference" in the society's journal, he applauded telepathy as the next step in human communications. "Telephone, telegraphs and words are too slow for this age; we must get something that is faster," he urged.[17] It may seem bizarre to compare our exploding, limitless experiences of cyberspace with past efforts to telegraph the dead or find proof of telepathy, a word coined in 1882 by its proponent F. W. H. Myers. But in both eras, inventions of science and technology tore down the constraints of time and space, firing the public imagination. Suddenly, spiritual and cerebral worlds, once considered the sole realms of religion and imagination, became cutting-edge scientific frontiers—and arenas for new forms of intimacy and communication. "Talking to the dead and talking on the phone both hold out the promise of previously unimaginable contact between people," notes historian Pamela Thurschwell.[18] Even the electric light, another invention that seemed to symbolize man's unlimited potential, was compared to a supernatural phenomenon.

Now exploratory forays into unseen worlds are burgeoning into a determined desire to increasingly *inhabit* new dimensions. In the past, mediums depended on spirit "guides" to describe the cricket pitches and "spirit universities" of the rather suburban-sounding Summerland, the area of afterlife most popularly described in spiritualist accounts.[19] Today, we don't need spirit guides to offer secondhand reports of other worlds. Rather, we spend increasing proportions of our lives in alluring netherworlds that we have constructed ourselves. "Unlike television, in which the audience is confined to the role of passive viewer, the game puts you into the action in the role of a god," a critic writes of the online alternative-world fantasy game Sims 2, "and not a distant, magnanimous god but one of those petty Roman gods who amuse themselves by toying with people."[20] Nearly eight million have joined Second Life, a kind of fantastical cloning of the world where politicians get interviewed, entrepreneurs make up to $200,000 annually in real dollars, and support groups bond, just as they would in "real life." Second life, second chance. Now "what once might have been called reverie," writes art historian Jonathan Crary, mostly takes place in mediated worlds of "preset rhythms, images, speeds, and circuits."[21] Are we perfecting "wired love," upgrading ourselves, grasping transcendence on earth through our screens?

Whether via mediums or cyber-games, past and current technologists share a stubborn certainty that utopian ideals of connection are within reach of explorers of these other realms. In its day, the telegraph was celebrated, even beyond its ability to shatter distance, as a device that, according to nineteenth-century British prime minister Lord Salisbury, "combined together almost at one moment . . . the opinions of the whole world."[22] Although leaders of the telephone industry at first erroneously assumed that, like the telegraph, the telephone was solely a business tool, industry literature by the 1920s promised that the invention would turn the country into a "nation of neighbors."[23] But perhaps few would have believed as passionately in the power of today's online social networks to connect humanity as Pierre Teilhard de Chardin, a French Jesuit and paleontologist (1881–1955). Teilhard, who experienced in his lifetime a dizzying parade of technological changes, fell afoul of the Catholic Church in his attempts to reconcile evolution with science and theology. His idea that the universe evolves seriously challenged the Church's view of a static planet with eternally fixed forms. As a result, the Church sent him into exile in China in 1926, forbade him from teaching theology or philosophy, and kept his greatest work, *The Phenomenon of Man*, from being published during his lifetime. Nevertheless, he made important contributions to paleontology and continued to write, developing in exile his concept of the "noosphere"—the idea that all human thought persists beyond its recorded forms and envelops the earth like another atmosphere. "As an effect of co-reflection, the human mind is continually rising up collectively because of the links forged by technology," wrote Teilhard, whose idea of the "thinking membrane" increasingly is being revived and cited.[24] A *New York Times* editorial lauded the "blogosphere" that is doubling every five months as the "global thought bubble of a single voluble species."[25] Network theorists, meanwhile, are exploring the idea that vastly different systems, such as the Web and the North American power grid, have a common dynamic and governing architecture, in other words, everything is ultimately connected.[26] Whether the noosphere strikes you as eccentric or brilliant, certainly Teilhard's vision presciently captures our mushrooming Web-based quilt of knowledge and relationships. If as nothing more than an analogy, the noosphere describes the expectation and, increasingly, the reality of our times.

Yet what ultimately happens when the simultaneity of yesterday becomes a quest for multiplicity, and pioneering explorations of unseen worlds become *inhabitations* of limitless new dimensions? There's a fine line, after all, between rich relations and meaningless hyperconnectedness, between abundance and chaos. How do we keep from getting lost in these mercurial and diffuse realms, where time for reflection and focus is increasingly lost as a valued part of life? To set the stage for a deeper appreciation of our own landscape, consider how our cultural views on attention have shifted in the past century. For when we try to shatter the earthly and *stretch toward* for the stars, we perceive the world differently. We shape our attention through our yearnings, choices, values, and blind spots, now and in the distant past.

"For what is even the greatest distance to modern man?" asked August Fuhrmann in 1905. "Nowadays it is easily bridged by railroads, air and steam ships, post and telegraph, so that it has almost ceased to be a factor in cultural life."[27]

Bringing faraway worlds to Europe's doorstep was, you could say, Fuhrmann's bread and butter. He was a German physicist and entrepreneur who invented a precursor of cinema and television that enthralled spectators from its launch in 1880 right up until the eve of World War II. The Kaiserpanorama, renamed the Weltpanorama after World War I, consisted of a large wooden cylinder with twenty-five individual binocularlike lenses. Peering in, spectators saw a show of rotating and seemingly three-dimensional slides of news events and exotic places, such as a Turkish harem or the pope's private apartments. For the average European of the day, the "stereopticon" show briefly captured slices of a world changing almost beyond belief. The invention proved so popular that the Kaiserpanorama was franchised across Europe, with two hundred and fifty machines in operation at one point. Since shows changed twice weekly, Fuhrmann sent his own troupe of photojournalists around the world to shoot news-making events and far-off places.

The Kaiserpanorama did more than bring a glimpse of distant lands to entranced spectators in the early days of photography. Fuhrmann's invention helped change how Victorians perceived the world, and from the seeds of this change have grown our own collagelike and mediated existence. "One great charm of the shows at the Kaiser Panorama was that it made no difference where in the series you started," recalls critic Walter Benjamin in a memoir of his childhood in Berlin. "Because the display went around in a circle, every picture would come around past the stop where you sat and peered through two little windows into its pale, tinted vistas."[28] The Kaiserpanorama offered a creaky, mechanized nineteenth-century virtual reality show, "a steady, constant flow of images whose narrative content was secondary to its promise of a vicarious visual experience of the real," observes art historian Angela Miller.[29] It was the era's *Sex and the City* and CNN crawl wrapped in one. This automated panorama became one of the first modern inventions to mediate and mechanically direct human attention.

It's no coincidence that the Kaiserpanorama proved spectacularly successful in an era that was fascinated by attention yet discomfited by the growing discovery that it could not always be controlled, understood, or protected. "Whatever its nature, [attention] is plainly the essential condition of the formation and development of mind," wrote Henry Maudsley in the early 1880s.[30] This intangible aspect of humanity inspired endless public debate and research by scientists, writers, and philosophers of the day. Scientists studied people's responses to electric stimuli or their own ability to concentrate their mind or vision. Psychophysicist Gustav Theodor Fechner noticed that when he held two different-colored objects in front of his eyes, only one color or the other could be seen after a time and that no matter how much he concentrated, he could not keep both in view.[31] Suddenly, attention was not the obedient phenomenon that he and his colleagues had once assumed, notes science historian Michael Hagner.

Remember that this was an era of rapid mobility, newly instantaneous communications, mystical, magical technologies—and broken norms. The world seemed at once flattened and fractured, shrunken yet imploded, mutable and subjective. In realms from art to science to politics, the comforting notion of a shared reality became passé. "With the tremendous acceleration of life, mind and eye have become accustomed to seeing and judging partially or inaccurately, and everyone is like the travelers who get to know a land and its people from a railway carriage," wrote Nietzsche, a leading proponent of a new philosophy called "perspectivism," which held that reality was a matter of interpretation.[32] In 1905, an obscure patent clerk named Albert Einstein published five papers that radically remapped conceptions of space and time. Most famously, his theory of relativity postulated that what we observe depends upon how we move through space. In parallel, artists and filmmakers were inspired by Paul Cezanne's pioneering depictions of subjects—such as his beloved Mont Sainte-Victoire—from multiple perspectives on a single canvas. Joyce portrayed a cacophonous world of clashing, blurring, overlapping realities. In *Ulysses*, "the real action takes place in a plurality of spaces, in a consciousness that leaps about the universe and mixes here and there in defiance of the ordered diagramming of cartographers," writes Stephen Kern in *The Culture of Time and Space: 1880–1918.*[33]

Perhaps inevitably, attention began to be seen as subjective and malleable, rather than the stable, controlled, and highly valued mental ability admired greatly during the ever-reasonable Enlightenment. Attention in Benjamin Franklin's day was the thinking man's obedient dog, the faithful servant of civilized man's efforts to overcome primitive superstitions and scientifically understand his world. (Physiologically, attention was thought to derive from a pooling of "nervous fluid" in a part of the brain.) To study insects or plants, eighteenth-century naturalists such as the Swiss entomologist Charles Bonnet undertook feats of observation so rigorous that they were ridiculed by satirists of the day. Bonnet once studied a single aphid from 5:30 a.m. to 11:00 p.m. for twenty-one days straight in order to learn about its reproductive cycle. And he took such detailed notes of his experiments that modern scientists often find it hard to determine which type of insect he was studying. "The part overwhelms the whole" in Bonnet's notebooks, writes historian Lorraine Daston.[34] Yet while Bonnet and his colleagues were undoubtedly extreme in their devotions, their views of attention as a sort of intellectual firepower, rigidly controlled by will and tenacity, reflected those of their era. "Around 1800, attention made us the masters of exploring ourselves and the world that surrounds us," observes Hagner. "Around 1900, the spaces between ourselves and the world was filled by apparatuses, instruments, technologies and all sorts of entertainment."[35]

By the nineteenth century, attention was no longer seen as a docile, willing companion, but a cerebral Hound of the Baskervilles—difficult to leash, frighteningly powerful, and vulnerable to control by external forces, both human and mechanical. "Now the risks, gaps and instabilities of attention were set upon stage," observes Hagner.[36] In particular, an outpouring of research into hypnotism explored the fine line between enrapt attention, sleep, and hypnotic trances, as well as the expansion of learning and memory that accompanies the intense focus of the hypnotized. The technique, although little understood, at first was lauded as a significant medical advance. Yet the idea that one's attention could be controlled by others, sometimes malevolently, proved deeply disturbing to the public and experts alike. George du Maurier's 1894 novel *Trilby*, one of the most popular novels of the decade in Britain and America, tells the story of a musician who hypnotizes and enslaves a young Bohemian girl, turning her into an opera star. In Bram Stoker's *Dracula*, the vampire not only sucks his victims' blood, he controls their minds through telepathy and hypnotism. "When my brain says 'Come!' to you, you shall cross land or sea to do my bidding," threatens the vampire before forcing the book's heroine, Mina, to drink his blood.[37] Ultimately, hypnosis became too closely associated with docility and a loss of control to be accepted by a public and scientific establishment struggling with the mutability of modern life. By the turn of the century, hypnotism increasingly was shunned by influential advocates such as Freud, writes Crary in *Suspensions of Perception: Attention, Spectacle and Modern Culture*. "There was an astonishing cultural reversal from the great heyday of hypnosis in the late 1880s," he writes, "when across Europe and North America it seemed a therapy that promised unlimited benefits, to the turn of the century, when it had become an embarrassment to its former advocates."[38]

Generations ago, people took their first forays into dreamy new worlds: from collagelike visions offered by the Kaiserpanorama and new spiritual playgrounds, to the unsettling embrace of hypnosis. The experiences were heady, even as the risks were real. In a world of shattering boundaries and assumptions, losing control was a peril ever-crouched on the other side of the front door. But now we splinter our focus and plunge into alternative realities so often that perhaps we are losing the ability and will to question the consequences. We begin to take for granted that attention is a kind of noisy, manipulated, and splintered battleground. As recently as the 1920s, silence was "the context of thought, conversation, and general existence," notes Joseph Urgo, author of *In the Age of Distraction*.[39] Now this space is "colonized," he writes. A realm of silence is as rare as a virgin forest and, to many of us, as eerie. "Inundated by perspectives, by lateral vistas of information that stretch endlessly in every direction, we no longer accept the possibility of assembling a complete picture," asserts literary critic Sven Birkerts. "We are experiencing in our time a loss of depth—a loss, that is, of the very paradigm of depth."[40]

Crary ends his book with a letter, dated September 22, 1907, that Freud sent to his family from Rome. Describing his evening, Freud tells how he stood in a crowd to watch a "lantern slide" show projected on the roof of a house in the Piazza Colonna. Freud doesn't describe the content of the slides but rather their effect. Every time he tries to pull away during an ad, "a certain tension in the attentive crowd" forces him to look back at the short films and still slides. "Until 9 p.m., I remain spellbound, then I begin to feel too lonely in the crowd, so I return to my room to write to you all," wrote Freud.[41] This vignette of the great thinker, far from home and entranced by a primitive precursor of the media that envelop us today, is somehow poignant. Freud is captivated, until he wakes from his automated reverie, feeling homesick and empty. Written more than a century ago, this brief sketch of one evening in Rome whispers to us of the skimming, fleeting, alluring world we now inhabit.

How did we come to the cusp of a new dark age? The seeds of our world can be traced to the first high-tech revolutions, when now quaint technologies shattered the confines of the local and the tedium of ancient routines. In the nineteenth century, people became untethered and empowered in striking new ways, and they began to experience the delights, quandaries, and questions that now confront us urgently and inescapably each day. As economist Jeremy Rifkin reminds us, "the great turning points in human history are often triggered by changing conceptions of space and time."[42] The telegraph introduced people to the thrill of simultaneity. Now we slip easily in and out of virtual worlds and multitask each other, wondering if our seemingly miraculous power to be in many places at once brings us closer or keeps us apart. At the same time, the fleeting glimmers of life at the other end of the telegraph wire or Ouija board have evolved into galaxies of otherness, real and imagined. Finally, the "pale vistas" of the Kaiserpanorama offered an enthralled public a first taste of a virtual, mechanized entertainment in an era captivated by the dark mysteries of attention. Increasingly, we sense that crucial aspects of our humanity—our ability to focus, be aware, and reason well—may be eroding, even as we surrender to the dreamlike joyride that this way of life offers. Now it's time to confront the challenges of our day. Does intimacy survive a seemingly limitless realm of infinite prospects? Can we bolster the quality of our life by split-screen living? How do lives of perpetual movement shape our attachments to each other and change our experience of place? Facing these challenges leads us first into the "new room" in the house, the virtual space where the lights are always on. But before we enter, let's give a last word to a largely forgotten nineteenth-century thinker who piercingly understood our times.

Albert Robida was a French cartoonist, illustrator, and novelist who surpassed even Jules Verne and H. G. Wells in describing with astonishing accuracy the day-to-day impact of technology on society. In his novels such as *The Twentieth Century* and *The Electric Life*, Robida predicted helicopters, giant-screen television, twenty-four-hour worldwide real-time news, video telephones, and test tube babies, along with biological warfare, environmental destruction, the domination of imagery over reality, and the speeding up of life. Equal parts utopianism, witty satire, and science fiction, Robida's works—unlike Verne's—explored the social impact of coming technologies, not simply the wonders of the hardware. At one point in *The Twentieth Century*, a hyper-busy father attends his daughter's wedding by telephone. In other scenes, two unsuspecting people are watched in their bedrooms by remote users of a Webcam-like "telephonoscope."[43]

His drawings and prose humorously eased people's anxieties about the era's rapid technological advances, according to Philippe Willems.[44] Yet Robida's message was clear: be warned about the consequences of unrestrained technological living. Born in 1848, the shy and myopic Robida rode a bicycle only once, detested automobiles, and refused to use the most wondrous technology of his day—the telephone. We could dismiss him as a curmudgeonly Luddite, but that would be missing the sparkling clarity of his vision. After writing a lighthearted short story titled "In 1965," he admitted in an interview that he didn't envy those who would live in our time. "Their every day will be caught in the wheels of a mechanized society," he said, "to the point where I wonder how they will find the time to enjoy the most simple pleasures we had at our disposal: silence, calm, solitude. Having never known them, they shall not be able to miss them. As for me, I do—and I pity them."[45] With trepidation, Albert Robida foresaw an age of distraction, and he depicted it well to an era that did not heed his warnings.

Chapter Two

FOCUS

E-mailing the Dead and
Other Forays into Virtual Living

Alan Edelson stood on the shore of the Michigan lake where his fourteen-year-old son, Zachary, had drowned. He flipped open his cell phone. Four years after the accident, Zach's name was still at the top of Edelson's phone list, just as his instant-messaging screen name "Snoopy1372" pops up when Edelson logs onto his home computer. "How do you push the delete button?" asked Edelson, a furniture store owner in suburban Detroit. "You don't."[1] He hung back from the water's edge, facing a sky darkened by storm clouds. An old man toting a fishing pole ambled off the beach, followed by a puppy. In the distance, a teenager reeled in his line and cast again. This was the first time that Edelson had been back to Walnut Lake since Zach died. It was a Fourth of July weekend, and Zach had just finished tubing with friends and had taken off his life jacket. A gust of wind sucked an inner tube out of the boat. A rope wrapped around his leg, dragging him overboard. He banged his head and sank.

It took three days to find his body in the lake's 80-foot, 45-degree waters. For three days and nights, Edelson stood on this spot, interrupting his vigil only to shower and sleep for a few hours. Hundreds of onlookers—mostly friends and family but a few gawkers—

congregated, periodically pulling out their cell phones to apprise others of the search. Edelson used his cell phone to tell his wife, Zach's stepmother, to come to the lake, to break the news to Zach's grandparents, and to try unsuccessfully to stop a TV news helicopter from hovering. On this weekday afternoon, the lake was calm, rimmed by still boats and empty lawns. But that seventy-two hours remains vivid for Edelson, a thoughtful man with thinning gray hair and a pale complexion. He is both tortured and soothed by memories of his popular, athletic youngest, the baby to two brothers and two stepsisters, now all moving out into the world on their own. Seeing virtual glimmers of Zach is comforting, insisted Edelson when we first met at a Starbucks in a strip mall near the lake. "He's there, he's still a part of you. Memories are what you have and you want to maintain them the best you can." Edelson reached for his iced coffee, wiping a bloodshot eye. He wore a pink Ralph Lauren oxford shirt with a smudge of soot on the front, chinos, and a broken look.

In talking about his loss, Edelson kept coming back to the importance of "moving on." That's why, after much debate, he and his ex-wife, Zach's mother, let the boy's e-mail account lapse, even though his friends kept writing to Zach—chatting, reminiscing, promising to remember him. That's why they decided not to create a memorial site on the Web, where "cemeteries" link page after page of tributes and missives to the dead, along with virtual flowers that never wilt. "People have to move on, too," said Edelson. "He had lots of friends. They have a life to live." Still, neither parent has touched Zach's rooms in their respective houses, nor have they removed his screen name and his listings in their cell phones. Those vestiges of their son, both physical and virtual, remain too real to ignore.

Not long ago, we thought of "virtual reality" as an all-encompassing, stomach-churning, and perhaps seizure-inducing category of games played with helmets and joysticks. A mental space trip, alluring but hardly mainstream. But now the virtual has infiltrated *real life*. Sure, thirty million play massively multiplayer online role-playing games (MMORPGs) such as Second Life and World of Warcraft, taking on alternative identities that reflect the multifaceted aspects of their complex selves.[2] But now the tentacled iterations of the virtual reach far beyond games. In a corner of the ether, a parent and child separated by thousands of miles share a bedtime book and a moment of togetherness by videoscreen as part of a court-ordered custody visit. An unattached college student lists a single friend as her Facebook "husband." The graffiti on a New York telephone pole prompts you to click on a Web poem. Two family members trade instant-message banter from separate rooms of the house or separate corners of the same room. College roommates feud the same way. A dictionary defines *real* as "actually existing" and *virtual* as "being so in practice (though not strictly) or in name." As sociologist Manuel Castells points out, reality long has been partly experienced virtually, that is, via symbols such as the alphabet or a currency. But nowadays, our communication systems offer us a world in which reality is so immersed in simulation and make believe that "appearances . . . become the experience."[3] The outcome is a culture of what Castells calls "real virtuality," or put more simply, the virtual as real. It's an online first love affair by two teenagers who have never looked into each other's eyes—and don't want to. It's a dead boy's AOL account that still gets mail.

All this can occur, as social scientist Sherry Turkle notes, because "we are all computer people now,"[4] and I would add, TV and cell phone people, too. To paraphrase Barry Wellman, only when technology is pervasive is its impact most felt.[5] We are now so surrounded by virtual experiences—as meeting place, blank canvas, augmentation of our materiality, vessel of our hopes and dreams and fears—that we've slipped into accepting its reality unthinkingly. We've minted a new currency of life and are using it alongside and as an extension of the old. I realized this with a jolt one day when I first attended one of the regular in-person meetings of a Web-based experts' group. I am a member of the group, which was organized by a think tank to explore the notion of virtual cooperation, and I had taken part in its online discussions but had never been to one of their face-to-face gatherings.

When the evening's host asked newcomers to stand and introduce themselves, I popped up, and she instantly looked uncomfortable. "Oh, yes, this is your first time *here*," she muttered. Only later did I realize my mistake: I considered myself new because I had never attended one of their flesh-and-blood meetings. But they accepted my virtual presence as time spent together. I had missed the point: the virtual counts. We take the faceless at face value now.

But what is this omnipresent space we increasingly inhabit? How does it change the nature of our attention to others? This New World is both more and less than an imitation of the physical, natural, and the earthly. Although most game developers desperately strive to make their fantasy worlds more "real," we can't and likely won't ever be able to clone our reality in fonts and pixels. That's obvious. What is not always apparent is how our time in this disembodied, alluring, liberating world changes us and especially influences our relations with each other. For this is no solo journey. Cyberspace, in science fiction writer William Gibson's words, is a "consensual hallucination,"[6] more so than any of the alternative worlds man has ever created—from Greek mythology to the layered slices of the medieval heavens to film or theater. We aren't just mulling and imagining our new realms. We live in them and thus carry out even our most intimate relationships "on-air." Understanding our virtual world—and its push-me-pull-you relationship with the physical—is a first step toward deciphering a time of distraction.

Someone asked me about you today,
It's been so long since anyone has done that.
It felt so good to talk about you,
To share my memories of you,
To simply say your name out loud.
They asked me if I minded talking about
What happened to you . . .
Or would it be painful to speak about it.
I told them I think of it everyday
And speaking about it helps me to release

The tormented thoughts whirling around in my head.
They said they never realized the pain would last this long.
T__, we're all thinking of you and miss you very much.[7]

A grieving man writes to his wife, more than a year after her death, on the Web memorial he created for her. Virtual cemeteries, such as the popular World Wide Cemetery, first appeared on the Internet in 1995, although individual memorials are now migrating out of the virtual graveyard and onto social networking and other sites. One man's virtual eulogy to his late father on a fishing hobbyists' site brought sixty-four responses in twenty-four hours.[8] Parents logging onto a deceased child's MySpace Web pages hours after the death to alert their son's or daughter's friends are surprised to see comments already posted, like notes left at a gravesite even before the burial.[9] While the Edelsons rejected the idea of cyber-mourning, thousands of others are attracted to the notion of creating such memorials as both public tributes to their lost loves and vessels of their grief.

Perhaps such virtual outpourings of emotion are not surprising at a time when earthly grief is considered to have a very limited shelf life. We mourn, yes, and always will, yet increasingly grief is treated as a private humiliation to be shunned, a step up from bodily wastes. Funerals are called celebrations of life, and once over, for many there is little evidence of mourning at home beyond a plate of cold leftovers and some wilting flowers. We certainly don't wear black for an extended time. Closure, quick as you can, is the goal. In her elegiac book on grief written after the death of her husband, poet Sandra Gilbert observes, "Just as we've relegated the dying to the social margins (hospitals, nursing homes, hospices), so too we've sequestered death's twins—grief and mourning—because they all too often constitute unnerving, in some cases indeed embarrassing reminders of the death whose ugly materiality we not only want to hide but also seek to flee."[10] Death and grief are just as potent, just as discomfiting as ever, and yet now they can be conveniently *sequestered*, packed away a few steps away from flesh-and-blood realities. One woman who set up a

Web memorial told researcher Pamela Roberts she was grateful for the supportive comments left there, as few would offer similar comfort in person. "For so many, it is just unbearable to come in direct contact with such raw pain and grief," explained the woman. The absence of communal rituals for mourning reflects a society, asserts David Moller, which increasingly "seeks to disengage suffering and . . . expressions of grief from the fabric of everyday society."[11]

Instead, the grief that cannot show its face in our postmodern culture is migrating to the Web, as are our age-old yearnings to communicate with those we have lost. These are dual compunctions: distance ourselves from the messy bits of life on the one hand, transcend the finality of death on the other. The grief-stricken have always talked to their dead, silently, out loud, in cemeteries and in letters—unmailed. Now, with a click and a keystroke, our messages are sent and posted for all to see. We may not get a "reply" as often as the nineteenth-century spiritualists who conjured up bells, whistles, ectoplasm, and coded missives from beyond the grave. Yet that only seems to heighten our sense of virtuality's mysterious potential and ironically even add credibility to our efforts to reach past the realm of the living. "I still believe that even though she's not the one on her MySpace page, that's a way I can reach out to her," Jenna Finke, twenty-three, says of her close friend, Deborah Lee Walker, twenty-three, who died in a Georgia car crash. Finke and other friends visit Walker's MySpace page—now a memorial site—daily, often to post messages to Walker.[12]

Gilbert said she had directed silent musings to her deceased husband occasionally, but once she "realized that he'd have an email address by now, I found myself standing before the glittering screen of the night sky and posting messages toward the virtual, though cryptic, space where I wish he'd be waiting to read them."[13] Strikingly, we don't even have to have known the dead in this life before trying to start up a conversation with them in the next. In one study, 35 to 42 percent of "guestbook" entries on memorial sites were written by strangers, and a quarter of those messages were addressed directly to the deceased.[14] Some were from visitors to nearby memorial sites, while others seem to be penned by cyberspace tourists peeking into yet another enticing corner of the virtual realm. After all, our technologies are at least ideally democratic. All we need in the age of science is the right machinery, the right calling plan, and we can do wonders. At the end of his life, Thomas Edison worked for years to build a machine that could get signals from the dead. "I have been at work for some time building an apparatus to see if it is possible for personalities which have left this earth to communicate with us," wrote Edison in his diary in 1920. In various entries, he fusses and blusters at his lack of success, saying with the right "scientific methods"—not, of course, the "childish nonsense" of the spiritualists—the deed will be done.[15] His staunch faith in his failed quest hangs in the air, like an unrequited dial tone. But we have moved on. Who needs mail from the deceased when we finally seem to have constructed the "apparatus" with which we can share their world, revel in its secrets, and then come home?

Many see in cyberspace nothing less than a new, spiritual heaven that is open to all who are computer literate, that is, "baptized," some observe. Cyberspace gives us the means to realize "a dream thousands of years old: the dream of transcending the physical world, fully alive, at will, to dwell in some Beyond—to be empowered or enlightened there, alone or with others, and to return," writes editor Michael Benedikt in the influential essay collection *Cyberspace: First Steps*.[16] Physicist and mathematician Margaret Wertheim argues in *The Pearly Gates of Cyberspace* that the inequity, fragmentation, and decline of American civilization inspires a deep desire to find spiritual meaning in cyberspace, just as late antique Rome's similar dissolution inspired numerous mystical religions. Promises of heavenly redemption made early Christianity especially popular.[17]

By its very nature, space is freedom, according to cultural geographer Yi-Fu Tuan, while place is akin to security and permanence. Hence, an inestimable attraction of the aptly named "cyberspace" is its breadth and size, its openness and impermanence. (Death, wrote Shakespeare, is just an "undiscovered country.")[18] "Spaciousness is closely associated with the sense of being free," Tuan writes. This, in turn, gives us "the ability to transcend the present condition, and this transcendence is most simply manifest as the elementary power to move"—and optimally, to go where you like. "Open space has no trodden paths and signposts," notes Tuan.[19] And yet space also connotes a threat, a sense of the peril of being lost or adrift. Consider Americans' fascination with the open road, the frontier, staking a claim, pursuing the space race, and you begin to understand that the virtual is not just as real to us as the physical; it is an alternative to the earthly. Computers are no longer "tools for the mind," observes Randal Walser, but "engines for new worlds of experience."[20]

But amid these wonders, doesn't home—the "boring old Earth," in the words of pioneering roboticist Hans Moravec—begin to pale in comparison with the virtual?[21] The grass *is* greener in the virtual realm. In this "enchanted garden" of endless delights, we leave behind a world mired in eternal decay, human foibles, and near-miss moments.[22] Our bodies become "meat" in William Gibson's argot. Our limitations begin to be unbearably confining. With the virtual at hand, we needn't suffer the discomfort of looking one another in the eye and fumbling for the right words to express our condolences. We needn't let a tear fall on a shoulder.

It was about seven o'clock on a late-spring evening, and Mae Cohen had just finished smothering a steak with a thick layer of chopped garlic from a hefty jar. "Is that enough?" she asked.[23] "Well, not any more than that," answered her mother, Beth, who was sitting at a worn kitchen table, pulling the skins from hot boiled potatoes for potato salad. The steak was a treat to mark Mae's return from her first year at a women's college a few hours from their house in a quiet Boston neighborhood. Her fourteen-year-old brother, Willie, was holed up in his bedroom squeezing in a last few minutes of gaming with two friends from school. ("It's medieval," he told me. "There's Diablo, the lord of terror. You're following him around so you can kill him.") The father of the family, Topher, was working late. Mae generously poured olive oil on a tray of asparagus, and the kitchen chatter turned to the year-long virtual romance between Willie's sixteen-year-old friend and a girl in Ohio.

"I think the kids like it better because it's very safe," said Beth, a computer software expert with a no-nonsense demeanor. "They can talk about how they are boyfriend and girlfriend, but the reality is all they're doing is talking on the phone" for hours each day. She doubted the pair would ever meet. Once, after they had persuaded their reluctant parents to let them see each other, the girl got cold feet. The boy planned a summer visit to relatives in Ohio but didn't want to meet the girl. When I wondered out loud what would happen if the two finally wound up in the same room, Mae was quick to answer. "Chances are, they wouldn't get along with each other." Meeting face to face would be "scary," said Mae, whose sweet, round face and dark brown hair were set off by a close-fitting embroidered cap. "The relationship could change." She seemed to speak with the certainty of experience.

After dinner, Mae told me about her favorite computer game, Alien Adoption Agency. To dress and train her aliens for battles, she earns "money" catching fish, whose scales, strangely enough, she sells to other players. Describing the game, she was more animated than at any other time in the evening. For a while in high school, she spent so much time immersed in the fantasy, which she plays with the graphics turned off so she can free up more space for chat, that she forced herself to quit for a while. When I asked just how much time Mae had spent back then dressing up aliens and sending them into battle, she deflected the question. "Every day," she said vaguely. I pressed her. "I don't remember," she concluded, ending my questioning with a shake of her head and a coy smile. She bit off a hunk from a chocolate bar. Recently, she adopted a new alien and acquired an online husband. "We've been married 106 days," she said proudly. Her mother, glancing through papers for Willie's new school orientation, said nothing.

Networks. Think of them as the ever-shifting constellations of relationships we inhabit on earth and in the "glittering screen of the night sky." On any given day or hour, we brush up against people from a dizzying spectrum of arenas—home, work, profession, school, sports, place of worship, chat rooms, games—without regard to their location. This facet of modern technologies—from cars to computers —has changed human life perhaps more than any other. Remember that for hundreds of years, settled peoples carried out their relationships mostly within bounded neighborhoods, largely "limited by their footpower in whom they could contact," writes Barry Wellman, a leading researcher of social networks. In the past century, our growing mobility meant that we obtained "support, sociability, information and a sense of belonging" as often as not from those living outside our neighborhoods and towns as from those within, according to Wellman. Yet most of society was still made up of "little boxes," with "precise boundaries for inclusion" and "interaction . . . in its place, one group at a time."[24] No one at the Elks meeting got a phone call from the foreman, unless perhaps the guy on the night shift had just dropped dead. No one, short of those desperate enough to send off for a mail order bride, had a far-off girlfriend they'd never met.

Now we live in a society dominated by what Wellman has dubbed "networked individualism." We can connect with almost anyone and at any time, but the connection is to the person and not to the place and largely to a slice of the person and not the whole. This means that Beth Cohen can teach online classes to students from around the country she'll likely never meet, hire a virtual team in Dallas without meeting its members until months after the fact, and keep in all kinds of flesh-and-blood and virtual contact with clients, coworkers, parents, and others—who are unlikely to know that she is also a mother, teacher, consultant, writer, or manager. Networked individualism means that Mae Cohen can "marry" a fellow online game player who may turn out to be a woman or just another kid.

This is "six degrees of separation" on overdrive, and the star of the show is the fast, placeless, and asynchronous medium of e-mail. Messages get sent to contacts of contacts, integrating social groups and increasing connectivity, although not predictably or uniformly, notes Wellman.[25] The upshot is a breathtaking breadth of relations. When danah boyd,* a sociology graduate student at the University of California at Berkeley, teased apart a five-year archive of e-mail messages saved by her friend Mike, she found that he had sent and received 80,941 messages with 15,537 unique people. Only 405 people received more than ten messages during that period from the twenty-four-year-old Mike, who works at an art installation firm in Boston. But taking his e-mail account as a yardstick of his connectivity, boyd found that Mike has ties to 662,078 people in the world, or 11.7 million people, if you count messages he sent or got with more than fifty recipients.[26] His social circles include family, high school, work, and college contacts mostly in New York, Boston, Texas, and California. It's hard to know if Mike is highly gregarious, but he doesn't seem unusually sociable by today's standards. Other studies have shown that while most Net users and nonusers have similar numbers of the closest or "core" ties, wired folks have bigger networks of weaker ties, such as acquaintances or work contacts. (As technology scholar Clay Shirky likes to say, technology won't ever change the number of people to whom you will donate a kidney.)[27] At one point, boyd herself had 278 friends linking her to 1.1 million others on the social networking site Friendster alone.[28]

The fear was that the Internet would turn us all turn into hermits, and maladjusted hermits at that. Of course, some starry-eyed early visionaries took Gibson's *Neuromancer* as the Bible of cyberpunk and the breathless pronouncements of Internet booster and poet John Perry Barlow—"I want to be able to completely interact with the consciousness that's trying to communicate with mine."[29]—as blueprints for future society. But no shortage of naysayers, in turn, warned that we'd all be sucked into the black hole of virtuality, alone in our rooms, pale and touch-starved, just as E. M. Forster depicted in his 1909 short story "The Machine Stops."[30] In the tale written to counter hyper-optimistic accounts of the future by H. G. Wells, people lead sedentary, virtual lives in underground cells controlled by a beloved Machine. "In the armchair there sits a swaddled lump of flesh—a woman, about five feet

*boyd spells her name with lowercase letters.

high, with a face as white as a fungus," Forster writes as the story opens. Forster's heroine, Vashti, and others live dronelike lives, avoiding movement, direct experience, and emotion. Ultimately, the machine begins to break down, plunging society into bewildered chaos.

There are glimmers of truth in Forster's melodramatic predictions, as is often the case with thoughtful science fiction. Robert Kraut's longitudinal studies of Internet users found that many initially showed symptoms of stress, depression, and loneliness and fewer face-to-face communications with friends and family after going online. After three years, these effects mostly dissipated, except for social introverts.[31] It's not uncommon to hear stories like that of Dave Peters, a Virginia teen who got so immersed in the Web in high school that he'd forget to eat dinner and let friendships shrivel. At college, he plummeted into a hermetic life of takeout and thirty-hour online role-playing sessions, until he met fellow players and realized with a jolt that they were mostly just depressed geeks, like himself. "It was a sort of power thing," he told the *Washington Post*. "I was a bigwig."[32] Peters is far from alone. Twenty percent of players of the MMORG game EverQuest say that they consider themselves denizens of the game who are just visiting earth.[33] It's hard to believe that they are earnest. Yet the Web, once it reached a certain vastness, perhaps attracts some who would have otherwise chosen to escape the pain of life through drink, drugs, or gambling. Chances are, a few always will find ways to disengage too completely. People are restless and naturally inclined to escapism, notes Tuan. "Fantasy that is shut off too long from external reality risks degenerating into a self-deluding hell—a hell that can nevertheless have an insidious appeal," he writes.[34]

But most of us haven't turned into pasty-faced loners with a social life limited to interactions with strangers on fantasy games. Instead, we carry out our relationships, old and new, weak and strong, gaming, and otherwise, on earth and increasingly in the shimmering landscape of the virtual. We juggle our twelve million connections in "portfolios of ties"[35] that inspire and require constant shifting, sorting, and prioritizing. The breadth of our relations, the ease and speed of our communications, and the heady freedom evoked by virtuality change how we pay attention to one another in many ways. And foremost, there are limitless options. Just as there is always another digital copy in the information age, there is always another of your millions of contacts on hand. As a result, "relations are more easily formed and abandoned," writes Wellman.[36] This has changed how we make and keep friends and jobs, as well as how we pursue love.

"This was a run of the mill date or maybe a notch better than that," says Greg, a twenty-something New Yorker whose love life was tracked by novelist Jennifer Egan for a magazine article on online dating.[37] He's describing a first date with a woman he met online, an evening that involved bar-hopping, grilled-cheese sandwiches, shooting pool, and sex at his apartment. "I liked her but not enough to merit fireworks. Given the seemingly endless selection, I get to be a little less forgiving," Greg concludes. Later, Greg meets a girl nicknamed Sam, and they begin a relationship that within a month or two looks a lot like what past generations would have called "going out." They spend most weekends together, meet each other's families, have sex, and trade e-mail love notes. But they insist they are just friends and keep their personal ads posted on dating sites, in Egan's words, "signaling to others while they sleep." In a world that values looser, freer, and often context-free relationships, Sam observes, "If it all goes terribly wrong, you can honestly just disappear."

Disappearing is easy. All you have to do is not answer, leaving the sender to question what happened, if they pause long enough to wonder. Sometimes you don't even mean to drop the ball, but you get busy, the e-mail grows stale, the impetus to answer fades away. I meant to set a date for dinner with my high school friend who moved back to New York, but our first few e-mail exchanges trailed off— eighteen months ago. I still wonder what happened to the college roommate who said she'd be in town and asked if we could get together, and then didn't answer when I said yes. Did she change her mind, find a better use of her time, remember that I'm a bore? Is my high school friend now asking herself the same questions? In modern society, we've always floated in and out of each other's lives, dropping well-intentioned white lies about "getting together soon." But our rampant connectivity leads to an epidemic of loose ends, fueled by both the allure of endless options and by miscues rampant in a medium steeped in both uninhibited intimacy and venom. We are ghosts moving in and out of each other's consciousness, often silently but sometimes with a shriek and a howl.

Nineteen-year-old Miguel de los Santos smiled and shrugged when he told me about the Mexican joke.[38] De los Santos, an easy-going fellow who emigrated from the Dominican Republic in seventh grade, sat behind a steel desk in an office at the Boston nonprofit where he worked part-time while attending community college. The cramped office in the IT department was chilly and lit only by a computer screen casting a glow on de los Santos's cherubic face. Wearing a lemon-yellow polo shirt, multiple gold chains, and his brown hair in long, curly ringlets, he looked like a misplaced surfer. His attitude of studied nonchalance, I sensed, comes in handy in an age when managing exponential social networks is almost a full-time job.

The joke he told to a friend of a friend while all three were in a chat room came from an irreverent television comic whose specialty is skewering fellow Latinos. When de los Santos repeated the joke, the friend's friend signed off and now refuses to speak to him. De los Santos's friend in turn chided him, saying he went too far, but he answered that if they'd heard the joke on TV, no one would have been offended. He insists that he's not bothered even when friends virtually turn their back on him like that. Some disappear permanently, others "close and block" him for months, then pop up one day, asking how he's doing. "I think it's funny. I think it's how the world has changed," said de los Santos, who similarly moves pals in and out of prominence on his "Top 8" friends list on the social networking site MySpace. He'll eject a friend who makes him angry or promote another to the list on his or her birthday. One high school acquaintance has won a permanent place on his Top 8 because she doesn't swear. "I've never heard her drop the F- or the B- words in five years of high school. I admire that, so I just gave her a spot," he said with a lordly air. "I keep the close people up there," he said of the Top 8, "so everybody knows who I care to talk to." One game developer calls children born into a mediated world "the undo generation," since they have a "start again" or "game over" mentality to everything in life.[39]

This kind of "reputation management" is a time-consuming but unavoidable practice in a world of unwieldy, amorphous networks. Singles Google prospective dates. Employers check out college recruits' Facebook pages. Friends check up on each other by scrutinizing changes in their blogs or networking sites. ("If I see something sketchy or weird, I call them and say, 'Is anything wrong?'" said de los Santos.) It's intriguing that reputation management demands and inspires skills that are strikingly similar to those needed for successful gaming: a sense of mastery, an ability to remain aloof, an instinct for ranking others. "The online community is like love," mused de los Santos as he ran his cursor over the photos of his Top 8 friends. "It's so mysterious." I peered over at the screen, where blurry snapshots of Maria, "Sweet Pea," Val, "Killa Kelvin," Whitney, and the others were posted in a neat formation, and I couldn't help thinking, what kind of love treats people like chess pieces?

At the end of the day, being networked is in some ways just a numbers game, a time squeeze that even the fertility of cyberspace cannot cure for us. Whether you call it being multifaceted or being fragmented, at its heart networked individualism is about diffusion. "Connectivity is up, cohesion is down," says Wellman.[40] For instance, Americans now average twenty-three "core" ties and twenty-seven less close "significant" ties, not including the thousands, perhaps millions of people that we, like Mike, are connected to in slighter, "liter" ways each day.[41] People with a core network of this size have 12 to 18 percent fewer contacts of all kinds—in-person, landline, cell phone—*except via e-mail* with their core ties than those with ten or fewer core ties. Similar changes occur in constellations of significant ties. Perhaps as a result, a quarter of Americans report having no close confidantes, double the number who reported such a degree of isolation in

1985. Today, most say they have just two people they can turn to for social support, compared with three in the mid-eighties.[42] A majority of teens, meanwhile, say they don't believe that the Internet can help them have better social lives, even while they spend nearly as much time socializing with their friends virtually as they do in person. (Ten hours a week on average in person outside school, versus nearly eight hours virtually.)[43] In a connected world that nurtures plentiful weak ties, we turn to thinner, faster, and instant ways to keep up with one another. De los Santos said that he, unlike many others, lists only those he has met in person on his MySpace page. But it's also clear that he rarely sees most of his Top 8 friends.

Undoubtedly, face-to-face contact remains what Turkle calls the "gold standard" of human relations. It's the oft-stated ideal of family life and, until we perfect cybersex, the goal of online daters. We still go to parties, bump into one another on the street, have lunch, catch a ball game, as we always will, all while lamenting that we don't do so more often. We're not loners. We're connecting with crackling ferocity. And yet in a world of real virtuality, is face-to-face nevertheless becoming an antiquated luxury we don't really have time for? Perhaps we're not disengaging completely from our old physical realities as much as gently choosing to downgrade them. When we do come together, especially with those we love, do we turn away?

A burly dad in a T-shirt and shorts stretches as he rises from the living room couch. He's just gotten home from work and is pausing to catch a few minutes of the evening news with his wife. "Call me when it's back on," hollers the dad as he heads to the family's cluttered home office, where his son, who looks about eight or nine years old, is playing a noisy computer game with two friends.

"Hey, dudes, what are you playing?" asks the dad.

The boys don't answer.

He tries again. "How was school?"

Silence.

"Hey, Bud," says the father, affectionately rubbing his son's head.

The boy, his eyes glued to the screen, speaks, "Hey, Dad, this is not working."

The father responds vaguely, then tries to jokingly engage the kids in conversation about the game. "What are you doing—are you cuckoo for Cocoa Puffs?" he asks, three times.

"No," the son finally answers, momentarily looking at his father for the first time.

The father leaves the room. "God forbid they should know how to do anything but play that," he mutters.

I watched this scene unfold on film on the laptop of Elinor Ochs, a linguistic anthropologist, MacArthur fellow, and head of UCLA's Center on the Everyday Lives of Families.[44] Since 2001, she has led the center in studying thirty-two racially and economically diverse Los Angeles–area working families in more detail than perhaps any households have been studied before. Archaeologists photographed their mountains of belongings. Psychologists studied the rise and fall of their stress hormones. Ethnographers with digital video cameras filmed the family members from dawn to bedtime for a week and analyzed the sixteen hundred hours of footage frame by frame. When I first called Ochs to ask about the families' use of technology, all she wanted to discuss was the moment at the end of the day when they reunited. Like the distracted son in the video, mostly the family members ignored one another. I was uninterested in such trivialities. Didn't everybody know that "Hi, Honey, I'm home!" is a bygone ritual?

Ochs is an intense woman whose black hair, small thin frame, and brisk demeanor are reminiscent of a mystical seer from some far-off tale. She has won worldwide recognition for helping create the field of language socialization—or how children are socialized through speech. When she speaks, she jumps from one seemingly unrelated topic to another, then unexpectedly sideswipes you with a thought that brilliantly ties her points together. On one of my visits with her, she talked about physicists concentrating on a problem, fundamentalists' trancelike praying, and children burrowing into their media. All are powerful forms of withdrawal and ways of organizing attention, she said. I began to understand. How and when we choose to move in and out of the reach of others shapes our relations. A greeting is a seemingly innocuous but crucial mark of our willingness to step forward, acknowledge another's presence, and, at day's end, mark a return to the fold. It is the opposite of withdrawal.

Alessandro Duranti, Ochs's amiable Italian husband, joined us for dinner at a crowded bistro one night near the UCLA campus, and we talked of their recent stay in Paris, music, and her findings. Duranti is an expert on the culture of jazz and a renowned linguistic anthropologist who has studied greetings in world culture for decades. Like funerals, greetings are rituals shared by all societies, usually occurring at the cusp or "near-boundary" of a social encounter as a way of acknowledging each other's presence.[45] (According to pioneering sociologist Erving Goffman, the establishment of copresence is a necessary first step toward crafting a mutual focus of attention in a social situation.)[46] Above all, greetings show who is worth recognizing socially, Duranti explains. "Just by greeting and not greeting, you're dividing up the world in terms of people who are worth talking to, and not worth talking to," he notes. "The people you don't greet are like shadows: they are there, they are not there." A greeting is a threshold moment in multiple ways.

In the families studied by Ochs's team, wives stopped what they were doing and welcomed home a returning spouse only a little more than a third of the time. Mostly, they were too irritable or busy to do so. Husbands did better, with more than half offering a positive greeting to a spouse. Children rarely greeted their fathers, who are mostly the last to return, and often didn't even look up when the dad reentered the house. "In half of these occasions, kids were absolutely distracted and didn't stop an activity to recognize a returning father," said Ochs. "They did not look up to stop even for a millisecond." When I talked to dozens of parents about their return at day's end, they laughed nervously, joked about their dog as the welcoming committee, then grew wistful and sometimes bitter, recounting the many nights they'd returned home to silence from a house full of busy kin.

Perhaps because we virtually check in with one another all day, the act of moving across a physical threshold naturally becomes devoid of meaning. In a placeless world, who needs to acknowledge the return to a location? Moreover, a boundaryless world means that coming home doesn't signal the end of the workday anymore than being on vacation is a time of pure relaxation or being under one roof marks the beginning of unadulterated family time. The physical and virtual worlds are always with us, singing a siren song of connection, distraction, and options. We rarely are completely present in one moment or for one another. Presence is something naked, permeable, and endlessly spliced. Add to this portrait of American home life the rise of networked individualism, and you begin to see why family members are not at the door, greeting one another. To cope and to keep up with our pulsing personal orbits, we live in worlds of our own making, grazing from separate menus, plugged into our own bedroom-based media centers, adhering to customized schedules. One in three employees is in contact with work weekly or more after hours.[47] Seventeen percent of the families in the UCLA study consistently ate dinner together. On weekdays, the parents and at least one child came together in a room just 16 percent of their time at home. True, hours together don't automatically translate into intimacy. But if we can't be bothered to keep coming together in the fullest, richest sense of the word, we lose the opportunity to form those deeper bonds. To relate, we need to have a "disposition to be vulnerable to interaction," said Ochs. Are we losing our willingness to wade down into the painful, soulful depths of human relations? "When you can have a face-to-face conversation, do you? When you're right in very close proximity, do you bother?" asked Ochs. "I'm afraid we're going to wake up and think, 'Oh my gosh, we could have been having a conversation.'"

Tammy Browning is proud of her efforts to hold her family together in a world that keeps tugging them apart.[48] A cheerleading coach in an affluent New Jersey suburb, she's used to putting a bright face on the unpleasant. She jokes that her husband Randy, a partner at a big accounting firm, travels so much that he must have a second wife and kids somewhere. With her three-year-old son in tow, she cheerily chauffeurs her fourteen- and sixteen-year-old daughters to modeling classes, voice lessons, dance competitions, parties, movies, sleep-overs, and school. A "health basket," stocked with granola bars, raisins, chocolate-covered nuts, and fruit, is kept in the kitchen so that the girls can grab "dinner" on their way to dance classes several nights a week. Gathering everyone in the same room is like a rare alignment of distant planets that's mostly outside of her control, something Tammy yearns for and keeps working toward. When her sixteen-year-old daughter Jordan told me "we're not a family, not like others . . . ," Tammy cut in: "We watch *Grey's Anatomy* as a family!" Then she laughed. "But we don't allow Randy to speak. He doesn't follow it, and he's like, 'Who's this? What's happening?'"

I was sitting on the Brownings' backyard poolside patio one sunny summer afternoon, talking to them about life in a wired family. Jordan, a pretty girl with wide eyes, said she tries to finish her homework before watching TV, unless one of her favorite shows is on. At the same time, she's usually playing music and instant messaging and texting friends. In fact, she and her sister Lindsay prefer to watch TV in their rooms, where they are most wired, even if their parents are watching the same show downstairs. Unplugging is so painful to Randy and the girls that when the family recently arrived for a wedding at a rural California ranch with no cell phone or Internet access, the three immediately drove to a local biker bar where they could get a signal. "It was a cozy kind of place, romantic," Tammy said of the ranch. "We're not woodsy people," interjected Jordan, who had twenty-two missed calls on her phone after six hours without a signal. As we talked, she and a friend named Dan busily took pictures of each other and texted other friends. Lindsay, who is also described by her dad as "surgically connected" to her cell phone, slipped out of the house and took a seat as Jordan started complaining that when she telephones her mom from upstairs, her mom doesn't pick up. After one or two rings, Jordan tries Tammy's cell, then Lindsay's. "They don't answer fast enough," said Jordan.

Moments after Tammy lamented how rarely the family eats together, Randy popped up from his basement home office to tell his wife that he had to go to the city for a client dinner. Before retreating sheepishly back into the house to change, Randy said, "I don't think people know how to socialize anymore, to sit down and come together, because there are so many alternatives. They like to be entertained, not socialize." Jordan protested. Later, on the half-hour ride to dance class, the sisters squabbled, wheedled snack money out of their mother, chatted about school, and clutched their cell phones, sending a steady stream of messages to their friends.

In updating our definitions of family for a new age, we've traded security for freedom and swapped the group for a loose network of kin. We don't seek suffocating ideals of obligatory togetherness, day by day or over our lifetimes. Yet all this changes the nature of family ties. "Kinship relations used to be taken for granted on the basis of trust, now trust has to be negotiated and bargained for, and commitment is as much an issue as in sexual relationships," says Anthony Giddens in *The Transformation of Intimacy*.[49] The recombinant family spawns what Ochs senses is a drift toward the voluntary—read: optional—nuclear family. Family ties, along with all relations, become a matter of choice. With technology lessening the need for direct human interdependence of all kinds, "association becomes a conscious act of will," says Michael Heim.[50] When the rise of the virtual disturbs our perception of what reality is, we suffer what philosopher Paul Virilio describes as a "loss of orientation" regarding "alterity (the other)."[51] It becomes easier to be lured away, to be tugged away and to deliberately turn away, countering the demands of always-on connectivity with powerful ways to filter out noise, toomuchness, and others. People are chess pieces and unwanted static. "With self-serve airport kiosks, ATMs, online grocery delivery services, clothing catalogues, restaurant reservations, and, of course, the ubiquitous iPod, it has become possible to filter almost every possible human interaction," writes Kate Zernike, adding that a minivan with a DVD is just a mobile way for parents to edit out their own children. "V-chip your television. V-chip your mom," she writes.[52] Leave nothing to chance.

But isn't chance at the heart of what constitutes the material, earthly, flesh-and-blood world? Contingency, or the "presence of the unforethinkable," is at the heart of earthly reality, according to philosopher Albert Borgmann.[53] Contingency doesn't make life random or meaningless, he says, adding that in ancient times *contingency* meant something like *consummation*. Rather, it attests to the "unsurpassable eloquence" of a reality that often asks us to confront what we cannot control or even understand. A face-to-face meeting, which demands a mutual reading of body language, emotion, and soul, is harder to fathom, and less predictable than a virtual encounter. But by losing the will to face one another, we are turning away from the messy, unpredictable, and *real* in life. Virtual pink slips, condolences, courtship, custody visits, lovers' break-ups—all relegate the hard parts of life to a thinner, indirect realm. More than seventy years ago, E. B. White could see this coming. Media will remake our world and "insist that we forget the primary and the near in favor of the secondary and the remote," he wrote.[54] Not too long into the Internet era, novelist Mark Slouka could see the truth of White's prediction. "What surprises us now increasingly," he observed, "is the shock of the real: the nakedness of face-to-face communication, the rough force of the natural world."[55] A mourner's tear, a lover's fallibility—these are becoming too much to bear. V-chip them out.

Ultimately, the virtual will probably not displace the physical. We know that we cannot leave our earthly selves behind. We live betwixt and between, on the edge of two worlds that are perpetually in a state of tension, says Sherry Turkle.[56] But what is frightening is that as Borgmann predicts, we seem to be degrading the earthly to a mere utility, a tool, a plane of reality we can hopscotch over when we want to avoid its hard, crude limitations. Against the virtual, Borgmann writes, "The actual world seems drab, poor and hard in comparison."[57] So we dip back into the here and now only when it suits us, sharing a meeting, room, or house with one another but staying tuned to other worlds and better opportunities. And in so doing, the virtual becomes the preferred reality. People long have found ways to disconnect from

each other, from sham marriages, insensitive parents, soured friendships, lost loves. But what if we choose to be satisfied by a steady diet
of glimpses, and little more? What if we, more often than not, opt to
turn and drift away? We become attentional wanderers, sated by the
mirage.

Before meeting Alan Edelson, I drove to Milford, a Michigan village
where the FBI had just ended two weeks digging up the Hidden
Dreams horse farm to try to find the remains of Jimmy Hoffa. The
alleged mobster disappeared in the area in 1975.[58] Milford, at least,
had a sense of humor about the $250,000 wild goose chase that
involved dozens of agents, geologists, archaeologists, cadaver-sniffing
dogs, and police. A bakery was selling thousands of cupcakes capped
with little green plastic hands sticking up through a heap of icing. A
Dairy Queen that I passed on the outskirts of town had a farewell message to the investigators: "Roses are Red; FBI is Blue; No bones for
them; Higher Taxes for U." On a quiet residential street, I parked in
the driveway of a large, white clapboard building and was met inside
by an astringent smell of cleaning fluid and a man who shimmered out
of nowhere to ask my business in a hushed tone. Posted by the door
was an Emily Dickinson poem.

> This is the hour of lead
> Remembered if outlived,
> As freezing persons recollect the snow—
> First chill, then stupor, then the letting go.

I had arrived at Lynch & Sons, Milford's leading funeral parlor. If
we are living at the edge between the earthly and the nether realms,
then Tom Lynch is the man who keeps an unflinching eye on our visits
to the borderlands. As the sign out front promises, he runs the family
business with his sons, but he is also a poet who writes of his Irish
roots, drunken past, failed marriage, and most often of all, of the stark
truths, both epic and quotidian, that he sees from behind an embalming

table in the village of Milford. He sees what happens when we leave this world and what happens to those who are left behind in a world that doesn't much come face to face with grief anymore. He sees a people who, more than ever, don't want to face death. As we talked in his cozy, book-lined office, he pulled down his latest work and read me a quote from Robert Pogue Harrison. We must choose, Lynch read, "an allegiance—either to the post-human, the virtual and the synthetic, or to the earth, the real and the dead in their humic densities."[59]

If there is one thing that Lynch has noticed after three decades in the business, it's that the corpse is increasingly the only person not welcome at the funeral. "What does it mean when we walk into a funeral parlor where we would have traditionally seen the dead laid out among their people, and now we go in and see a golf bag-shaped cremation urn and next to it, the life-sized golf bag," demanded Lynch, a small man with a salt-and-pepper scrub beard and piercing blue eyes. The funeral, he said, has become a retail event cum life celebration that fails to "transport us to the border of a changed reality." The focus on the trappings distracts us from the reality of death, a realm we can't control, gloss over, predict, or ultimately avoid. Death is the epitome of the earthy and the messy but also the epitome of the real. "Eventually, the only reason we pay attention to the dead is that death ennobles life, makes life more meaningful," said Lynch. "Where death is meaningless, life becomes meaningless." What will it be, Lynch demanded, and I fidgeted uneasily in my silky soft chair in front of his enormous desk. For a minute, he was no longer the kindly undertaker, but the daunting gatekeeper, book in hand. I thought of W. H. Auden's poem, "Musée des Beaux Arts," which muses about our detachment to others' suffering and concludes with an homage to Pieter Bruegel's portrait of a farmer indifferent to Icarus's fall from the sky.[60]

> . . . how everything turns away
> Quite leisurely from the disaster; the ploughman may
> Have heard the splash, the forsaken cry,
> But for him it was not an important failure . . ."

"So which will it be," asked Lynch, "the golf bag urn (read: the post-human, virtual, synthetic), or the humis, the ground, graveyard, village, nation, place or faith—the nitty-gritty real earth in which human roots link the present to the past and future?"

The body at the threshold at the end of the day. The body at the threshold at the end of life. Will we turn away?

Chapter Three

JUDGMENT

Of Molly's Gaze and Taylor's Watch: Why More Is Less in a Split-Screen World

Molly was busy. A cherubic, dark-haired fourteen-month-old still unsteady on her feet, she hung onto a bookcase with one hand and doggedly yanked toys off the shelves. One, two, three brightly colored plastic blocks dropped to the floor. A teddy bear got a fierce hug before being hurled aside. Then abruptly, she stood stock still and swiveled her head toward a big television set at one end of the room, entranced by the image of a singing, swaying Baby Elmo. "She's being pulled away," whispered Dan Anderson, a psychology professor who was videotaping Molly and her mother from an adjoining room of his laboratory. "What's happening is that she's being pulled away by the TV all the time, rather than making a behavioral decision to watch TV."[1] As Anderson and two graduate students observed through an enormous one-way wall mirror and two video monitors, Molly stood entranced for a few seconds, took a step toward the screen and tumbled over. Her young mother, sitting on the floor nearby, turned her attention from the television in time to catch Molly before her head hit the floor. Anderson didn't react. He was tuned to the back-and-forth in the room: Molly turning from the toy to the TV to her mother; her mother watching her baby, but mostly the video, which was being developed by Sesame Street for the exploding under-two market. This was rich fodder for a man who's spent his life studying children's attention to television.

A congenial University of Massachusetts professor with a melodic voice, Anderson resembles a character in a fairy tale—perhaps the gentle wizard who shows the lost child the way home. First in viewers' homes and now in his lab in a working-class neighborhood of Springfield, Massachusetts, he studies heart rate, eye tracking, and an array of other measures to understand what happens when we watch television. People aren't as glued to the tube as they might think, Anderson has found. On average, both children and adults look at and away from a set up to one hundred and fifty times an hour.[2] Only if a look lasts fifteen seconds or longer are we likely to watch for up to ten minutes at a stretch—a phenomenon called "attentional inertia."[3] When a show is either too fast and stimulating or too calm and slow, our attention slips away. Television attracts us because its content can challenge our cognition. But foremost, its quick cuts and rapid imagery are designed to keep tugging at our natural inclination to orient toward the shiny, the bright, the mobile—whatever's eye-catching in our environment. It's ingenious: entertainment that hooks us by appealing to our very instincts for survival. This is why very young viewers like Molly are entranced by the plethora of new "educational" shows and DVDs aimed at them, even though they understand little and likely learn little from this fare.[4] Push and pull, back and forth, television is in essence an interruption machine, the most powerful attention slicer yet invented. Just step into the room with the enticing glow, and life changes.

This was the intriguing discovery that Anderson made while exploring the gaze of the tiniest watchers, the final frontier of TV viewership. In all the years that he and others sought to probe the question of how much we attend to television, no one thought to ask how television changed off-screen life during an on-air moment. (The point of most such research, after all, was to measure the watching, the more of it the better, in the industry's view.) But Anderson and his research team recently discovered that television influences family life even when kids don't seem to be watching. When a game show is on, children ages one to three play with toys for half the amount of time and show up to 25 percent less focus in their play than they do when the TV is off.[5] In other words, they exhibit key characteristics—abbreviated and less focused play—of attention-deficient children.[6] They begin to look like junior multitaskers, moving from toy to toy, forgetting what they were doing when they were interrupted by an interesting snippet of the show. Not surprisingly, parents in turn are distracted, interacting 20 percent less with their kids and relating passively—"That's nice, dear" or "Don't bother me, I'm watching TV"—when they do. Consider that more than half of children ages eight to eighteen live in homes where the TV is on most of the time.[7] Factor in the screens in the doctor's office, airport, elevator, classroom, backseat—and don't forget that many, if not most, are splintered by the wiggling, blinking crawl. Then, zoom out and remember that television is just one element in a daily deluge of split focus. Wherever Molly's gaze falls, wherever she turns, whomever she talks to, she'll likely experience divided attention. She's being groomed for a multitasking, interrupt-driven world. And she doesn't need Elmo to teach her that.

If the virtual gives us a limitless array of alternative spaces to inhabit, then multitasking seems to hand us a new way to reap time. Cyberspace allowed us to conquer distance and, seemingly, the limitations of our earthly selves. It has broken down the doors of perception. Now, we're adopting split focus as a cognitive booster rocket, the upgrade we need to survive in our multilayered new spaces. How else can we cope with an era of unprecedented simultaneity, a place we've hurtled into without any "way of getting our bearings," as Marshall McLuhan noted in 1967.[8] Multitasking is the answer, the sword in the stone. Why not do two (or more) things per moment when before you would have done one? "It's a multitasking world out there. Your news should be the same. CNN Pipeline—multiple simultaneous newstreams straight to your desktop." I am watching this ad on a huge screen at the Detroit airport one May evening after my flight home is canceled. Travelers all around me move restlessly between PDA, iPod, laptop, cell phone, and ubiquitous TV screens. "CNN Pipeline," the ad concludes. "Ride your world." Rev up your engines, Molly, it's a big universe out there.

Now working parents spend a quarter of their waking hours multitasking.[9] Grafted to our cell phones, we drive like drunks; even if it kills us, we get in that call. Instant-messaging's disjointed, pause-button flavor makes it the perfect multitasking communications medium. More than half of instant-message users say they always Web surf, watch TV, talk on the phone, or play computer games while IM'ing.[10] Teens say, duh, *that's* the attraction: face to face is better, but with IM, you get more done![11] Joichi Ito's "hecklebot," which publicly displays the "back-channel chat" or wireless banter of conference attendees, may be just the pipe dream of a subversive venture capitalist for now, but it captures the tenor of the attentional turf wars erupting in meeting rooms, conference symposia, and college classes.[12] "Did he really say that?" instant-messages an audience member to fellow IM'ers in the room. "Wow! He did," someone responds.[13] This parallel channel adds a new layer to the surfing and e-mail checking already rife at live-time events and makes the dualing dialogues in Daniel Ho's Technology Play seem more truth than fiction. Bosses, speakers, and professors respond with threats and electronic blackouts to wrest people's primary focus back to the front of the room. Audiences ignore them, asserting the right to split their focus. Are these just bumps along the road to progress? Can we timesplice our way to unlimited productivity? Certainly, the disjunction between TV news anchors and the crawl "captures the way we live now: faster than ever, wishing we had eyes in the back of our heads," notes media critic Caryn James.[14] The inventor of the Kaiserpanorama put it more simply. In an age of technological wonders, merely traveling to far-off places won't be enough, wrote August Fuhrmann. Next, we'll want to penetrate the unknown, do the impossible—instantaneously. "The more we have, the more we want," he wrote.[15]

Another day in the lab, and this time I was the baby. Sitting in a cramped booth in a basement laboratory at the University of Michigan in Ann Arbor, my head was cradled in a chinrest and capped by a headset. My right hand rested on a set of four metal keys. On the table in front of me, two eyeball-shaped videocams sat atop a computer screen, monitoring me as I struggled to correctly respond to beeps, squeaks, and colored words—red, blue, yellow, green—appearing on the screen. "Beep," I heard and tried to recall if that was supposed to be sound one, two, or three. The lone word *red* appeared on the screen, and I thankfully remembered to press the corresponding pinkie finger key. Two practice rounds and then paired tones and colors flew at me simultaneously, even though I seemed to sense just one and then, after a long pause, the other. The colors I could handle, but sometimes I didn't even hear the tones. I pictured Adam and Jonathan, the two graduate students in the next booth, rolling their eyes as they ran this test taker through her paces. I pressed on, trying to concentrate. It felt like gritting my teeth, except in my brain.

David Meyer, head of the University of Michigan's Brain, Cognition, and Action Lab, was my guide that day to the burgeoning realm of cognitive neuroscience research into multitasking.[16] Considered by many of his peers to be one of the greatest experimental psychologists of our time, Meyer looks more like an outdoorsman than a brilliant scientist. Lanky and tall, he has a chiseled face and a down-home way of talking, with faint traces of a Kentucky accent. Blessed with the ability to translate brain science into plain English, he's become a media darling in recent years, the one to call for a quote on the latest multitasking research. He's more than generous with his time and willing to endure the interruptions of press calls. He's also a driven man.

Dressed in a faded T-shirt and blue jeans, he'd dragged himself into the lab this stifling May morning despite a painful stomach ailment. Now sixty-four, he's made it a point in recent years, even at a cost to his time for other research, of warning anyone who will listen about the dangers of multitasking. It's an unrecognized scourge, he believes, akin to cigarette smoking a generation ago. Is he riding a hobbyhorse, perhaps overreacting a trifle? Certainly, his call-a-spade-a-spade demeanor has raised eyebrows in the button-down scientific community. He writes lengthy scientific papers and speeches when snappy, four-page reports are increasingly in fashion. He refuses to lard his work with superficial pandering citations to big names in the field. At the same time, Meyer is a renaissance scientist, respected for his achievements in areas from computational modeling of the brain to "semantic priming"—or the automatic spread of mental and neural activity in response to processing the meaning of words. Is he a provocative maverick devoted to a peripheral pet cause or a prophetic visionary who can help save us from ourselves?

Certainly, it's ironic that Meyer is best known in the public sphere for his work in an area of study that long was a backwater in attention research. By the time Wilhelm Wundt established the first psychology lab at the University of Leipzig in 1879, a generation of scientists had spent years studying how humans perceive the world, especially visually.[17] The discovery that we *interpret* daily life via our senses, not just digest it objectively, paved the way for endless attempts to rigorously measure how a human responds to environmental stimuli and to what extent the waters of perception are influenced by the forces of "memory, desire, will, anticipation and immediate experience," as delineated by art historian Jonathan Crary.[18] Part of the vision of William James stems from the fact that he never underestimated the complexity of such processes. Yet however crucial and enigmatic, the "input-output" transactions that fascinated early psychological researchers entail only one slice of the pie of cognition.

It wasn't until after World War II that scientists began to see that studying how we switch mental gears, especially under pressure, can illuminate the higher workings of the mind. Arthur T. Jersild had carried out the first systematic study of task-switching in 1927 for his dissertation by timing students solving long lists of similar or differing math problems. Then he abandoned the topic, never to return.[19] Later, postwar British scientists began tackling task switching as part of their groundbreaking research into higher cognitive processing. A parallel line of research probed dual tasking, or our limited capacity to carry out two tasks literally at the same time.[20] By the 1990s, an explosion of research into multitasking had ignited, inspired by the work of Alan Allport, Gordon Logan, David Meyer, Stephen Monsell, and Harold Pashler, among others—and by the demands of life today. It's not a coincidence that such research has blossomed in an era when human work has become increasingly wedded to the rhythms of the most complex, intelligent machines ever to appear on earth. (In part, Meyer is known for his work with computer scientist and cognitive psychologist David Kieras, using computers to model the brain's cognitive architecture, including the "mechanics" of task switching.)[21] The question of how our brains compare with artificial information processors, and how well they can keep up, underlies much of our fascination with multitasking. We're all air traffic controllers now.

Back in the booth, I tackled a different experiment, this one measuring the speed at which I could alternate between two complex visual activities. Although the first experiment tested my ability to respond to simultaneous stimuli, both effectively measure task switching, for we can do very few things exactly at the same time. Reading e-mail while talking on the phone actually involves reading *and then* chatting, chatting and *then* reading. Cell phoning while driving demands similar cognitive switching. In this second test, when a zero popped up in one of four spaces in a line on the screen, I was to press a corresponding key with my finger. A zero in the first spot meant that I should press my index finger key. A zero in the second place prompted my second finger, and so on. Tap, tap. I got it. Easy. That was the compatible round. Next, I was supposed to hit keys that did not correspond with the zeros in the old lineup. When I saw a zero at the end of the line, I was to strike my index finger key. There was a pattern, but I barely grasped it before I had to begin alternating between compatible and incompatible cues, depending on whether the zeros were green or red. I panicked, blindly hitting any key in the harder, incompatible rounds and was thoroughly relieved when it ended. Yes, William James, there's a whole lot more going on here than just simple inputs and outputs. My brief cerebral tussle in the test lab, in fact, neatly exemplifies the age-old, inescapable tug of war we experience each waking minute of our life as we struggle to stay tuned to and yet make sense of our world.

To understand multitasking, first consider the lowly neuron, especially in the three-dozen regions of the brain that deal with vision—arguably the most crucial attentional sense. They lead something of a dog-eat-dog life. Neurons in the retina initially transmit an object's simple, restricted characteristics, such as its color and position, while more lofty neurons in the cortex and other areas code the object's complex or abstract features, such as its meaning. (Is this a face or a toaster, my neighbor or my mother?) This hierarchy of neurons must work in concert, firing up a coordinated "perceptual coherence field" in scientist Steven Yantis's words, to meaningfully represent the object in the brain.[22] But with so much to process and so little time, multiple neuron groups often compete to represent sensory information to the brain for possible subsequent encoding into memory. What is the key to making meaning from this jumble? Attention. Paying attention, whether deliberately or involuntarily, highlights one coherence field and suppresses activity from "losing" neuron groups, forcing our perception of the object they are representing to fade away. Attention is so crucial to how we see the world that people with damage to areas usually in the brain's right parietal lobe—a region key to certain forms of attention—can completely fail to notice objects in their view even though their vision is perfect. Such patients with "visual neglect" will eat only the food on the left side of their plate or dress just the left side of their bodies.[23] They literally have blind spots, no-go zones for their attention. And yet even for those of us with healthy brains, focus itself creates a kind of blindness. When we shine our attentional spotlight on an object, the rest of the scene doesn't go blank, but its suppression is truly dramatic. "The intuition that we open our eyes and see all that is before us has long been known to be an illusion," notes Yantis.[24]

We aren't built, however, to tune out life. Our survival lies in the tricky push and pull between focusing and thus drawing meaning from the world, and staying alert to changes in our environment. This is the real tug of war. As much as we try to focus on pursuing our goals, at

heart we are biased to remain alert to shifts—especially abrupt ones—
in our environment. Babies and children are especially at the mercy of
their environments, since it takes many years and much training for
them to develop the brain capacity to carry out complex, goal-oriented
behaviors, including multitasking. Older toddlers whose mothers constantly direct them—splicing and controlling the focus of their attention—show damaged goal-setting and independence skills a year
later.[25] Even as adults, our "top-down" goal-oriented powers of attention constantly grapple with our essentially more powerful "bottom-
up," stimulus-driven networks.[26] Pausing along the trail to consider
whether a plant was edible, our ancestors had to tune out their environment long enough to assess the would-be food. But they had to be better
wired to almost unthinkingly notice the panther in the tree above—or
they would have died out rapidly. We are born to be interrupt-driven, to
give in Linda Stone's term "continuous partial attention"[27] to our
environment, and we must painstakingly learn and keep striving to
retain the ever-difficult art of focus. Otherwise, in a sense, we cede control to the environment, argues physicist Alan Lightman in an essay
titled "The World Is Too Much with Me." After realizing that gradually
and unconsciously he had subdivided his day "into small and smaller
units of 'efficient time use,'" he realized that he was losing his capacity
to dream, imagine, question, explore, and, in effect, nurture an inner
self. He was, in a sense, becoming a "prisoner of the world."[28]

When we multitask, we are like swimmers diving into a state of
focus, resurfacing to switch gears or reassess the environment, then
diving again to resume focus. This is a speeded-up version of the push
and pull we do all day. But no matter how practiced we are at either of
the tasks we are undertaking, the back and forth produces "switch
costs," as the brain takes time to change goals, remember the rules
needed for the new task, and block out cognitive interference from the
previous, still-vivid activity.[29] "Training can help overcome some of
the inefficiencies by giving you more optimal strategies for multitasking," says Meyer, "but except in rare circumstances, you can train
until you're blue in the face and you'd never be as good as if you just

focused on one thing at a time. Period. That's the bottom line." Moreover, the more complex the tasks, the steeper the switch costs. When I had to consider both tones and colors in the first experiment, I began responding almost twice as slowly to the easier color tasks as I also tried to concentrate on getting the hard-to-hear tones right. Perhaps recalling which finger key corresponded to which color word, or in Meyer's words "rule activation," inhibited my performance. Perhaps my brain was slowed by "passive proactive interference," in other words, it was still tied up with the work of distinguishing the tones, a sticky business for someone whose hearing has been eroded by years of city living. Similar trade-offs occurred during the second experiment. I slowed down in doing the easy compatible work, while trying like mad to speed up my responses to the infuriatingly illogical second round of zeros. Predictably, the accuracy of my responses often suffered. These lab rat exercises and millisecond "costs" may seem abstract. Sure, an instant of inattentional blindness or a delayed reaction in noticing a darting child makes an enormous difference in a car flying down the road. The split-focus moment literally may result in shattered lives. But scale up and out of the lab and ask, how much does this matter off the road or away from the radar screen? Is multitasking as much of a scourge as Meyer believes?

Perhaps the cumulative, fractional "switch costs," the cognitive profit-loss columns of our split-screen life, are not the only problem. These are inefficiencies, surely a danger in some circumstances and a sin in this capitalist society, which we undoubtedly will try to shave away by sharpening our multitasking skills. More importantly, perhaps in this time-splicing era we're missing something immeasurable, something that nevertheless was very much with me as I struggled to act like a good test monkey in Meyer's lab. How do we switch gears in a complex environment? Talk to a cognitive neuroscientist or an experimental psychologist such as Meyer and chances are, within minutes, he or she will stress the limitations of our highest form of attention—the executive system that directs judgment, planning, and self-control. Executive attention is a precious commodity. Relying on multitasking as a way of life, we chop up our opportunities and abilities to make big-picture sense of the world and pursue our long-term goals. In the name of efficiency, we are diluting some of the essential qualities that make us human.

Frederick W. Taylor, a Philadelphia aristocrat who came of age in America's industrial heyday, probably would have enjoyed Dave Meyer's company. Although born to great wealth in 1856, Taylor as a youth apprenticed himself as a factory hand and was so committed to making his own way in life that he refused his inherited share in his father's considerable fortune. Photographs of the up-and-coming factory manager depict a thin, determined man with a steely gaze. Taylor was a reformer whose plain talk often ruffled feathers. He was also the father of our relentless drive to "work smarter," the man whose theories, according to management guru Peter Drucker, have had as much of an impact on the world as those of Marx or Freud.[30] Our propensity to multitask is something of a legacy from this myopic engineer whose ideas were both famous and vilified in his lifetime.

Taylor was reared by a puritanical Quaker mother whose adamant belief in "work, drill and discipline" helped foster his love of mathematics, mechanical inventions, and, above all, precision. A born athlete despite poor eyesight, he was so obsessed with refining the rules and techniques of the many sports he played that he reportedly took all the fun out of neighborhood games as a child. He spent hours working out the best strokes and strategies in the infamously precise game of croquet. After he took up golf at age forty, he tinkered with new types of clubs, soil mixtures, and fairway grasses to improve the game. His efforts often paid off: he was the 1881 national doubles tennis champion and the handicap champion at the Philadelphia Country Club in 1902, 1903, and 1905.[31] When he turned his attention to improving the realm of work, there were no halfway solutions for Taylor.

At the time that Taylor began working as an apprentice machinist at the Enterprise Hydraulic Works of Philadelphia in the late 1870s, industrial manufacturing was still a rough and unpredictable enterprise. In those pre-assembly line days, efficiency was not rewarded. Workers routinely set a molasseslike pace on the job, both because they were paid by the hour and feared that they'd be fired if they consistently finished their work early. Foremen, the only species of manager, were loath to dirty their hands alongside workers they saw as lazy and expendable. "Productivity" and "motivation" were unknown words, and not surprisingly, perpetually tense labor relations regularly exploded into violence. No one had actually studied work with a view to how it could be improved—until Taylor.[32] Over the course of two decades while holding a string of positions in both labor and management, Taylor honed a complex, brilliant, multifaceted web of solutions that ultimately inspired the fields of industrial psychology, human resources, performance evaluation, quality control, and management studies, and changed the way people work globally.[33] Even Vladimir Lenin was a convert. "We should try out every scientific and progressive suggestion of the Taylor system," the communist revolutionary was quoted as saying in *Pravda* in 1918, three years after Taylor's death at age fifty-nine.[34] Taylor's 1911 book, *The Principles of Scientific Management*, ultimately became the most popular business book in the first half of the century, and his ideas helped seed Japan's economic advances in the twentieth century.[35] Yet during Taylor's lifetime, labor leaders kept their distance from his ideas and most managers branded him a "dangerous radical" and "troublemaker."

Why were his ideas so influential and yet so controversial? At the heart of his reforms lies a simple concept: the fragmentation of work into measurable and ideally interchangeable pieces. Taylor's first and most powerful innovation was using a stopwatch to time workers' motions, then cutting out unnecessary or overly taxing movements. "Where others were content with the total time a job took, Taylor broke the job into its component parts, analyzed each part, and reconstructed the job as it *should* be done, more efficiently and with less fatigue," writes management professor Daniel Wren.[36] Taylor also espoused incentive wages, labor-management cooperation, and work tailored to individual skills. But these innovations pale in comparison to the legacy of his stopwatch management. Taylor did for work what the concurrent inventions of the cinema and photography did for vision and perspective. Stopping, starting, speeding up, freezing a moment: to discover these new sensations was to control time, and thus the shape of progress. This was the era of Muybridge's high-stepping horses, captured frame by frame in the performance of their unwitting ballet, and Picasso's shattered cubist still-lifes, which dissected and then reconstructed daily life. In film, the camera's "lowerings and liftings, its interruptions and isolations, its extensions and accelerations, its enlargements and reductions" not only rendered the world more precisely but revealed underlying structures and patterns we were not conscious of, observes Walter Benjamin in his classic essay "The Work of Art in the Age of Mechanical Reproduction."[37] No wonder the poet and journalist Charles Leland trumpeted the idea of "nimbleness of attention" for this new age. In 1891, he wrote that children need such skills of focus to be able to remember and retrieve rapidly accumulated perceptions, as if they were fishing mental snapshots from a keepsake box.[38] This was thinking as data processing, notes Stephen Arata, with efficiency "driving the whole enterprise, [and] attention its fossil fuel."[39]

But dissecting and distilling the components of work in order to turn a factory into a "well-designed, smoothly running machine" had a chilling outcome.[40] Taylor effectively treated man as an interchangeable part of the industrial machine. This was his "blind spot," concludes Drucker, and the source of labor's hostility to him. Even today, many workers are loath to cooperate with the more than fifty Taylor-inspired time measurement techniques in use in factories.[41] (His promotion of high wages, sharp criticism of management, and caustic demeanor in turn repelled employers.) "Taylor destroyed the romance of work," writes Drucker, who nevertheless admired Taylor's principles. "Instead of a noble 'skill' [work became] a series of simple motions."[42] Taylor furthermore advocated a strict separation of planning *from* doing in the workplace. "We do not let [workers] think. We do the thinking," he said in a 1907 lecture. Another of his sayings: "In the past man has been first.

In the future, the system must be first."[43] (No wonder Lenin loved him.) Taylor wasn't heartless and he didn't invent but rather exemplified the ideals of his age and the values that still shape our work world. He likely would not have approved of multitasking, as he believed firmly in having workers do one task at a time. Yet his thinking shadows our lives. He was the high priest of our unshaken veneration of productivity and arguably the grandfather of a work style that fragments tasks—at times to Chaplinesque extremes. Taylor's one-time protégé Frank Gilbreth, who turned his household into a time management laboratory, once found that he could cut forty-four seconds from his morning shaving routine by using two razors. But Gilbreth, who also set up back-to-back tonsillectomies for his dozen children to save time, abandoned double-handed shaving after finding that he had to spend two minutes applying bandages to all the cuts. "It was the lost two minutes that bothered him and not the cuts," recounts Wren.[44] No moment was left untouched in the pursuit of efficiency.

Now most of us are information-age workers, doers and planners both. No men stand over us with stopwatches. We are our own efficiency experts, relentlessly driving ourselves to do more, ever faster. This relentless quest for productivity drives the nascent but rapidly burgeoning field of "interruption science," which involves the study of the pivot point of multitasking. For multitasking is essentially the juggling of interruptions, the moment when we choose to or are driven to switch from one task to another. And so to dissect and map these moments of broken time is to shed light on how we live today. What emerges, in the jargon of leading interruption scientist Gloria Mark, is a portrait of "work fragmentation." We spend a great deal of our days trying to piece our thoughts and our projects back together, and the result is often an accumulation of broken pieces with a raggedy coherence all its own. After studying workers at two West Coast high-tech firms for more than one thousand hours over the course of a year, Mark sifted the data—and was appalled. The fragmentation of work life, she says, was "far worse than I could ever have imagined."[45]

Workers on average spend just eleven minutes on a project before switching to another, and while focusing on a project, typically change tasks every three minutes, Mark's research shows.[46] For example, employees might work on a budget project for eleven minutes but flip between related e-mails, Web surfing, and phone calls during that time. This isn't necessarily all bad. Modern life does demand nimble perception, as Leland observed, and interruptions often usher in a needed break, a bit of useful information, or a eureka thought. Yet as well as coping with a high number of interruptions, workers have a tough time getting back on track once they are disconnected. Unlike in psychology labs, where test takers are cued to return to a previous task, workers have to retrieve a lost trail of work or thought themselves when interrupted. Once distracted, we take about twenty-five minutes to return to an interrupted task and usually plunge into two other work projects in the interim, Mark found.[47] This is partly because it's difficult to remember cognitive threads in a complex, ever-shifting environment and partly because of the nature of the information we are juggling today. The meaning of a panther's presence is readily apparent in a glance. But a ping or a beep doesn't actually tell much about the nature of the information. "It is difficult to know whether an e-mail message is worth interrupting your work for unless you open and read it—at which point you have, of course, interrupted yourself," notes science writer Clive Thompson. "Our software tools were essentially designed to compete with one another for our attention, like needy toddlers."[48] Even brief interruptions can be as disruptive as lengthy, if they involve tasks that are either complex in nature or similar to the original work (thus muddying recall of the main work), Donald Broadbent has found.[49] In total, interruptions take up 2.1 hours of an average knowledge worker's day and cost the US economy $588 billion a year, one research firm estimated.[50] Workers find the constant hunt for the lost thread "very detrimental," Mark reports dryly.

To Mark's surprise, she and other researchers discovered that nearly 45 percent of workplace interruptions are self-initiated. (And when workers interrupt themselves, they take slightly longer to resume their original work—about twenty-eight minutes on aver-age.)[51] In other words, we are not only naturally catering to the "needy toddler" of a high-tech environment, but we are training ourselves to flit from task to task. "To perform an office job today, it seems, your attention must skip like a stone across water all day long, touching down only periodically," writes Thompson.[52] This is Taylor's real legacy to the information age. Like the voice of the disciplining parent still reverberating in our heads, Taylor's admonitions to chop up work to do it faster have become a part of the American work ethic. In burying the mystique of craftsmanship, Taylor destroyed the age-old idea of a human-paced flow of work. The word *multitasking* originated in computer lingo for multiple parallel processing. Now, as Mark and others have found, we bounce between an average of twelve different work projects and spend just twenty seconds per open window,[53] our split focus testifying to our efforts to pattern ourselves after the machine. To the Taylorization of work, we've added the "booster rocket" of multitasking. How can we find our way through the resulting landscape of lost threads and loose ends?

In the 1880s, as Taylor began arduously carving out his principles of management on the noisy shop floor of Philadelphia's Midvale Steel Company, William Morris was already pondering the fate of a culture of fragmentation. Head of England's most famous design firm, Morris was a successful businessman, revered poet, weaver, furniture maker, interior decorator, novelist, translator, printer, painter, political activist, and more—often simultaneously. A restless, nervous, fero-ciously energetic Victorian who was prone to bursts of temper, Morris —nicknamed Topsy—once described himself in a letter as "too busy . . . in all ways."[54] His doctor ascribed his death at age sixty-two to his "simply being William Morris, and having done more work than ten men."[55] Poet and Morris disciple Edward Carpenter wrote, "His chief recreation was only another kind of work."[56] Even his instantly rec-ognizable designs, crowded with twisting tendrils and intricate blooms, display an "obsessive activity," notes novelist Peter Ackroyd.

"He forever filled every available space"—and certainly every moment of the day.[57] His contemporaries often remarked upon his penchant for multitasking; he'd compose poetry while weaving, sketch wallpaper while conversing. If he were alive today, would this type-A multitasker fit in well with the restless crowd at the Detroit airport—checking e-mail, wolfing down hotdogs, barking into phones, scanning CNN headlines? Is this why Morris is remembered, his designs are sought after and exhibits of his work are jammed?

Perhaps there is an enduring allure in the breadth of his talents and the scope of his energy, but Morris speaks to us today because, for all his seemingly disparate activities, he was *not* diffused. He was a visionary who could see far beyond the shop floor of his small stained glass, furniture, textiles, and wallpaper business to future implications of the industrial era for work and for humanity. All of his endeavors have a consistency of thought: they were a response to a civilization that was evolving, in his view, into a "counting house on top of a cinder-heap." (At times, he resorted to endearingly tiny protests against the modern world: he once sat upon his top hat to protest "the market.")[58] All his life, he worked to infuse an industrialized world with a sense of the beauty of craftsmanship and the wisdom of the past. Both practical and idealistic, Morris was prodigious in his output, but, unlike Taylor, output was not his goal.

In founding his firm in 1861 with a group of Oxford friends, Morris set out to establish an arts and crafts collective that would produce high-quality goods for both upper- and middle-class homes. In this life's work, Morris found inspiration in nature and in all things medieval because in that preindustrial era, craftsmen had retained a "sense of the total endeavor," notes Peter Stansky.[59] Ackroyd adds that Morris "believed that eternity resides in the unique detail rather than in some Newtonian world of infinitely reproducible parts." Yet the short, corpulent, and forever unkempt Morris was no machine-busting Luddite. He commuted by rail for a time, although he loathed trains. And to make an affordable line of goods for the middle classes, Morris used mechanical looms in his workshop. "As a condition of life, production by machinery is altogether an evil; as an instrument for forcing on us better conditions of life, it has been, and for some time yet will be, indispensable," wrote Morris in an 1888 essay titled "The Revival of Handicraft."[60] Most importantly, he argued that happiness will elude us "if we hand over the whole responsibility of the details of our daily lives to machines and their drivers."

For Morris, taking pleasure in one's work was the essential ingredient of a good life, and its destruction the main crime perpetrated by the factory system. Work should involve both mind and body, and especially an unfettered mind. In his 1890 utopian novel *News from Nowhere*, the protagonist William Guest learns to cultivate a higher form of attention to the world as he tours a future world based on fulfilling work for all.[61] This state of mind, which Morris calls "repose amidst energy," is akin to our notion of being "in the zone" or in a state of "flow," when doing and thinking are melded and we lose ourselves within a challenge that ultimately raises us to a higher state of being. To Morris, this type of attention was also a political statement, a response to the industrial economy's increasing efforts to cultivate rigidly high levels of concentration in workers and schoolchildren, according to Arata.[62] In 1877, the designer refused an opportunity to become a professor of poetry at Oxford, partly because he opposed the growing institutionalization of reading, which began to resemble a manufacturing skill, complete with time management mandates. Even his habitual multitasking was not done in the spirit of inflexible concentration. Unlike the task-switching endemic today, Morris set his own pace for conversation, weaving, or writing poetry. (And he usually juggled a physical and cerebral activity, which scientists say is more doable than multitasking that uses similar areas of the brain, such as speaking on the phone while reading e-mail.) Most of all, you get the sense from reading contemporary accounts that for Morris, multitasking was a kind of sport. He himself admitted that he practically dashed off his poems. Perhaps this is one reason why, although he was known in his day as a great poet, his poetry seems to us placid and facile—as Ackroyd notes, "Tennyson rewritten by a town crier."

Epics aside, it was Morris's ability to dream that fueled his capacity to create and to envision a society that could still revel in the unbroken moment and the unfettered imagination. For this reason, both George Bernard Shaw and William Butler Yeats called Morris a prophet.[63] His genius lay in the totality of his efforts and his ability to see clearly the steep costs of an age that values the material and the efficient above the immeasurable and the human. He could see coming what Dave Meyer knows well and what Alan Lightman woke up to— the slow death of reverie. Perhaps this overarching vision is why Morris's busy designs are oddly calming; they evoke a natural landscape, holistic and integrated, never overwhelming. His talent for envisioning better worlds also fueled his unrelenting optimism. For despite the impossibility of realizing his broad schemes for societal change, despite his frustrations and tirades, Morris never stopped delighting in life and knew that it is often in its wondrous details that contentment can be found, Yeats recalled in an 1896 posthumous reminiscence. "In all his art one notices nothing more constant than the way in which it heaps up, and often in the midst of tragedy, little details of happiness," wrote Yeats in the *Bookman*.[64] In contrast, while Taylor's ideas would indeed come to change the world, the Philadelphia engineer died an unhappy and misunderstood man, battling for a fuller execution of his principles of scientific management.

If the Victorians had William Morris as the sometime answer to their ills, we have Mary Czerwinski, an energetic Microsoft researcher who designs a kind of high-tech "wallpaper" to better our age. Czerwinski is the manager of the Visualization and Interaction Research Group in the company's thought ghetto, Microsoft Research Labs. She originally wrote her dissertation on task switching, spent time helping NASA determine how best to interrupt busy astronauts, and now develops ways for computer users to cure that uncertainty rap—the necessity to unveil an interruption to size up its importance—mainly by bringing our information into the open, so to speak. Czerwinski and Gary Starkweather, inventor of the laser printer, are developing a forty-two-inch computer screen so that workers can see their projects, files, or Web pages all at once. That's three-feet-plus of LCD sensurround, a geek's heaven. Moreover, within this big-screen universe, Czerwinski and her team are figuring out new ways to make interruptions instantly visible. A program called Scalable Fabric offers a peripheral zone where minimized but still visible windows are color-coded and wired to signal shifts in their status. A new e-mail, for example, might glow green in a partly visible in-box. Another project creates a round, radar screen–type window at the side of the screen, where floating dots represent pertinent information.[65] Czerwinski is, in effect, decorating the walls of cyberspace with our thoughts, plans, conversation, and ideas. Can the "pensieve"—the misty fountain that conjures up the stored memories of Harry Potter's sage headmaster, Albus Dumbledore—be far behind?

Working memory is the Achilles' heel of multitasking, and so is the focus of Czerwinski's work. The "lost thread" syndrome that bedevils multitaskers stems from the fact that we have a remarkably limited cerebral storehouse for information used in the daily tasks of life. (Even a wizard, it seems, is a forgetful creature.) "Out of sight, out of mind" is all too true, mainly because, for survival purposes, we need to have only the most pertinent current information on our mind's front burner. Our working memory is a bit like a digital news crawl slithering across Times Square: constantly updated, never more than a snippet, no looking back. Nearly a half-century ago, memory researchers Margaret and Lloyd Peterson found that people forget unrelated letters and words within just a few seconds once they are distracted or pulled away to another task.[66] In his classic 1956 paper "The Magical Number Seven Plus or Minus Two," George Miller hypothesized that people could hold about seven pieces of information, such as a telephone number, in their short-term verbal working memory. The seven bits, however, could also be made up of "chunks" of longer, more complex, related information pieces, noted Miller, a founder of cognitive psychology. Recent evidence, in fact, suggests that Miller was overly optimistic and that people can hold between one and four chunks of information in mind.[67] Moreover, when your working memory is full, you are *more* likely to be distracted. This is one reason why viewers remember 10 percent fewer facts related to a news story when the screen is cluttered by a crawl.[68]

When I first talked to Czerwinski by telephone, she began the conference call by teasing a PR person on the line for failing to send out an advance reminder of the appointment.[69] "When I don't get a meeting reminder, you might as well hang it up," she said. To Czerwinski, the solution to the "lost thread" syndrome is simple: use technology to augment our memories. Of course, this is not entirely new. The alphabet, Post-It note, PDA, and now Czerwinski's innovations represent a long line of human efforts to bolster our working memories. But while multiple streams of color-coded, blinking, at-a-glance reminders will undoubtedly jog our memories, they run the risk of doing so by snowing us even more, Czerwinski admits. Bigger screens lead to lost cursors, more open windows, time-consuming hunts for the right information, and "more complex multitasking behavior," she observes. I would add that simultaneous data streams flatten content, making prioritization all the harder. The crawl, for instance, effectively puts a grade-B headline on a par with a top news story read by the anchor. Thirty shifting color-coded screen windows vying for our attention make trivia bleed into top-priority work. "Better task management mechanisms become a necessity," is Czerwinski's crisp conclusion. In other words, we need computers that sense when we are busy and then decide when and how to interrupt us. The digital gatekeeper will provide the fix.

And that's exactly the vein of research being mined by bevies of scientists around the country. "It's ridiculous that my own computer can't figure out whether I'm in front of it, but a public toilet can," says Roel Vertegaal of Queen's University in Ontario, referring to automatic flushers. Vertegaal is developing a desktop gadget—shaped like a black cat with bulging eyes—that puts through calls if a worker makes eye contact with it. Ignored, the "eyePROXY" channels the interruption to voice mail. An MIT prototype mousepad heats up to catch your attention, a ploy we might grow to loathe on a hot summer day. IBM software is up to 87 percent accurate in tracking conversations, keystrokes, and other computer activity to assess a person's interruptability.[70] The king of the mind-reading computer ware, however, is Czerwinski's colleague and close collaborator, Eric Horvitz. For nearly a decade, he's been building artificial intelligence platforms that study you—your e-mail or telephone habits, how much time you spend in silence, even the urgency of your messages. "If we could just give our computers and phones some understanding of the limits of human attention and memory, it would make them seem a lot more thoughtful and courteous," says Horvitz of his latest prototype, aptly named "BusyBody."[71] Artificial intelligence pioneer John McCarthy has another adjective to describe such programming: annoying. "I feel that [an attentive interface] would end up training me," says McCarthy, a professor emeritus at Stanford.[72] Long before "attentive-user interfaces" were born, French philosopher Paul Virilio had similar qualms about the unacknowledged power of the personal computer itself, which he dubbed a "vision machine" because, he said, it paves the way for the "automation of perception."[73] Recall David Byrne's impish observation that PowerPoint "tells you how to think."

Is hitching ourselves to the machine the answer? Will increasingly intelligent computers allow us to overcome our limitations of memory and attention and enable us to multitask better and faster in a Taylor-inspired hunt for ever-greater heights of efficiency? "Maybe it's our human nature to squeeze this extra bit of productivity out of ourselves, or perhaps it's our curious nature, 'can we do more?'" asks Czerwinski. Or are we turning over "the whole responsibility of the details of our daily lives to machines and their drivers," as Morris feared, and beginning to outsource our capacity for sense-making to the computer? To value a split-focus life augmented by the machine is above all to squeeze out potential time and space for reflection, which is the real sword in the stone needed to thrive in a complex, ever-shifting new world. To breed children for a world of split focus is to raise generations who will have ceded cognitive control of their days. Children today, asserts educator Jane Healy, need to learn to respond to the pace of the world but also to reason and problem solve within this new era. "Perhaps most importantly, they need to learn what it feels like to be in charge of one's own brain, actively pursuing a mental or physical trail, inhibiting responses to the lure of distractions," writes Healy.[74]

Ironically, multitasking researcher Arthur Jersild foresaw this dilemma generations ago. Inspired by Taylor's and other time management and piecework theories, Jersild quietly published his pioneering dissertation on task switching. Then he went on to become a developmental psychologist known for urging schools to foster self-awareness in children. His views were unusual. At the time, educators didn't consider children self-perceptive and, in any case, they felt that emotional issues were the purview of the family. In a 1967 oral history given upon his retirement from Columbia, Jersild argued that children must be taught to "see themselves as capable, if they are; to be aware of their strengths; to try themselves out if they seem hesitant; . . . to prove to themselves that they have certain powers."[75] Jersild, the sixth of ten children of a strict Midwestern Danish immigrant minister, was kindly, sensitive, and doggedly self-sufficient himself. At age fourteen, while boarding as a farmhand in South Dakota in 1916, he stood in the cornfield he was weeding, raised a fist to the sky, and vowed, "I am getting out of here!" By the end of his career, he had come full circle from his early concerns about how fast workers could do piecework to worrying that the education system placed too high a premium on speed and not enough on reflection. "It is essential for conceptual thought that a person give himself time to size up a situation, check the immediate impulse to act, and take in what's there," said Jersild. "Listening is part of it, but contemplation and reflection would go deeper." The apt name of his dissertation was *Mental Set and Shift*.

Depending too heavily on multitasking to navigate a complex environment and on technology as our guide carries a final risk: the derailing of the painstaking work of adding to our storehouses of knowledge. That's because anything that we want to learn must be entered into our long-term memory stores, cognitive work that can take days and even months to accomplish. Attention helps us understand and make sense of the world and is crucial as a first step to creating memory. But more than simply attending is necessary. "We must also process it at an abstract, schematic, conceptual level," note researchers Scott Brown and Fergus Craik. This involves both rote repetition and "elaborative rehearsal," or meaningfully relating it to other information, preferably not too quickly.[76] To build memory is to construct a treasure trove of experience, wisdom, and pertinent information. If attention makes us human, then long-term memory makes each of us an individual. Without it, we are faceless, hence our morbid fascination with amnesia of all kinds. Building these stores of memory takes time and will. When we divide our attention while trying to encode or retrieve memories, we do so about as well as if we were drunk or sleep deprived. In the opening scene of Alan Lightman's chilling novel *The Diagnosis*, successful executive Bill Chalmers loses his memory while taking the subway to work.[77] Rapidly, he disintegrates into a lost soul who recalls only his company's motto: "The Maximum Information in the Minimum Time." He regains his memory only to contract a mysterious, numbing illness that ultimately reveals the emptiness of his life. Like the alienated protagonists of film and literature from *The Matrix* to *The Magic Mountain*, Chalmers is a prisoner of the modern world. A culture of divided attention fuels more than perpetual searching for lost threads and loose ends. It stokes a culture of forgetting, the marker of a dark age. It fuels a mental shift of which we are not even aware. That's what we're unwittingly teaching baby Molly as her budding gaze meets the world.

Meyer's voice rose. He was hollering. We were back in his university office, and he was attacking the "myth" that we can operate at top speeds on multiple tasks as well as if we were doing one at a time. "That's ridiculous," he practically shouted. "That's ludicrous!" Like William Morris, Meyer often multitasks, reading the *New York Times*, chatting with his granddaughter, watching television all at once. Indeed, as we talked on the telephone one weekend morning, he interrupted me once or twice to read me the Wimbledon scores off the TV. But that's for fun, he insisted. "I'm getting little bits and pieces of information as a result of engaging in each of these tasks, but the depth of what I'm getting, the quality of my understanding is nowhere as good as if I was just concentrating on one thing," he said. "That's the bottom line." Once during our many conversations, he sadly confided the reason why he first began to be willing to speak up about multitasking: his seventeen-year-old son, Timothy, was broadsided and killed by a distracted driver who ran a red light one night in 1995. Spreading the word about the inefficiencies of multitasking is a little bit of "payback" for Tim's death. But that's only part of the story. Now Meyer speaks out about the costs of multitasking because he's convinced that it exemplifies a head-down, tunnel vision way of life that values materialism over happiness, productivity over insight and compassion. He's optimistic, taking a Darwinian view that people eventually will realize that multitasking's larger costs outweigh its benefits. But first Meyer believes that the few in the know will have to do a whole lot of hollering before we recognize the magnitude of the problem and begin to change. So he's raising his voice, as much as he can.

Chapter 4

AWARENESS

Portable Clocks and Little Black Boxes: The Sticking Point of Mobility

Tucked in a sunny corner of a noisy corporate cafeteria, I was lunching with Regina Lewis, the head of consumer insight for Dunkin' Brands, makers of Dunkin' Donuts, Baskin-Robbins ice cream, and Togo's sandwich chains. We foraged from a multitiered bakery cart stacked with platters of cruller-shaped pastries, round quiche-like breakfast pizzas, and thin, crusty, grilled-cheese sandwiches called "flats" that Lewis seemed to adore—all inventions being beta-tested in select stores. And as she nibbled through several sandwiches, Lewis told me what Americans want to eat these days. First of all, anything drippy, crumbling, or spillable is out, since so many people eat in their car and don't want to stain their clothes. "These are not pizzas where stuff is going to fall off everywhere and you're going to take a bite and half of it is going to be in your lap," said Lewis, a petite blonde dressed in a pale green sweater that matched her eyes.[1] She was right. The quichelike "southwest-style" pizza that I was eating—its copious cheese embedded with flakes of green and red pepper and bits of corn—likely wouldn't break apart if dropped from a considerable height. "Our R&D team has been able to devise pizzas that you can literally go like this with," said Lewis. With a proud flourish, she gestured as if shaking a pizza upside down.

We need handheld, bite-size, and dripless food because we are eating on the run—all day long. Nearly half of Americans say they eat most meals away from home or on the go.[2] Forty percent of our food budgets are spent eating out, compared with a quarter in 1990.[3] Twenty-five percent of restaurant meals are ordered from the car, up from 15 percent in 1988.[4] Moreover, this mobile foraging, propelled by time droughts, constant accessibility to food, and relentless travel, is redefining the notion of a meal. Americans report that 20 percent of their "meals" aren't breakfast, lunch, or dinner.[5] That's because snacking, generally frowned on a generation ago, is a norm, while meals tend to happen when and where we can fit them in. The two have converged, and we've become a nation of rampant "sn'eal-ers," or, if you prefer, devotees of the "mea-ck." For many, the day is one, long graze. Lewis, who is single and often goes for weeks at a time with nothing but water and milk in her apartment, recalled her shock when a man in a focus group described a late-afternoon snack of two McDonald's cheeseburgers. Lewis pressed him, "Was that a meal or a snack?" but he insisted that was just a snack. "Like, that was a snack!" laughed Lewis, who is strikingly feline in her girlish manner, velvety voice, and aloof air. She laughs often, perpetually amazed by today's eating habits, yet simultaneously seems to hold herself back, guarded, ready to move on at a moment's notice. "The fuel that people use to get them through the day now is different. They don't seem to be satisfied with just a bag of potato chips or crackers."

Cheeseburgers aside, Lewis empathizes with what she calls "perpetual lunchbag" living, the idea that food is a daylong endeavor that demands constant vigilance and strategizing. "When I go to our franchisee meetings now," Lewis related breathlessly, "I've been in a car, I've run to catch a plane, they haven't served food on the plane, I'm running to get to the meeting and so often, I'm thinking, 'I hope there's food there. I hope there's food there,' because I can't eat when I like, even though I do carry food around with me." For thousands of years, humans have honed a skill called "optimal foraging" that involves using the least energy possible to extract the most food from their habitat.[6] Now, we face a postmodern wrinkle: we're foraging as often as not on unfamiliar ground. We're perpetual tourists, grabbing and eating, searching and moving. We revel in our sheer ability to move and yet we are somehow surprised and even distressed by the burden of it all.

The grandchild of a West Virginia miner, Lewis's relentless curiosity sprung her from a rural cage. She went to college in Chicago, business school in New York; got a doctorate in communications at Duke; owned her own consulting business in San Francisco; and has wound up for the past three years living outside Boston, in an apartment not far from the sprawling corporate headquarters where she works. She thinks nothing of restaurant-hopping in Maine with New York friends one weekend, then jetting to Italy the next to run a marathon with a friend from London. Just back from a focus group in Nashville, she recalled listening to people who keep up a frenetic daily pace, all while insisting they are still laid-back. "They want to hold onto a simpler, easier time, but the truth is, when you watch their behavior, America is on the go," said Lewis, who has no such illusions about the tempo of life. When I asked her if it's hard to become rooted in Canton, a burgeoning suburb rimmed with industrial parks, she looked at me quizzically. "I don't have any interest in being rooted here, honestly," she said slowly. "My safety net is made up of individuals. So while I need to have associations with time and space, I can be anywhere." And she laughed.

The last key to decoding the landscape of distraction is movement. Just as the virtual shatters our conceptions of *space* and simultaneity redefines our notions of *time*, so a life of perpetual movement reshapes our relationship to *place* and what it means to be in the world. Grazing, browsing, surfing, hoteling, hot-desking, travel-soccer, dashboard dining—the very vocabulary of our daily lives denotes an adoration of motion. We scorn the idea of stopping, equating the local and the immobile with "social deprivation and degradation," writes sociologist Zygmunt Bauman.[7] We increasingly shape our lives around portability—of contacts, knowledge, entertainment, nourishment, and selves—and this changes how we sense and pay attention to the world. Do we notice the world around us or hurry through our days? Are we detached from ourselves and this earth? In the land of distraction, attention is restless and untethered, a wandering thing. And life is fluid and rootless. After my lunch with Lewis, I hopped on an Acela train back to New York, riding south as the sun set on the rails I'd traversed north that morning. Just for fun, I had deliberately told a sprinkling of people throughout the day—a cabbie, security guard, businessman on the train, marketing manager—that I was traveling five hundred miles for a lunch date. No one was the least bit impressed. However near or far they were journeying, they, too, were on the move and were anesthetized to our wondrous conquering of distance. In New Haven, a bevy of women who'd been attending a conference boarded the train, toting neat red gingham paper lunch boxes. One by one, they soon began poking inside, picking at their sandwiches and chips as we hurtled into the dusk of another day.

We are an infamously mobile nation. That is one of the cornerstones of the American dream and a root cause of our equally infamous stores of optimism. Our short history, immigrant roots, and pioneer spirit feed our yearning for novelty and freedom, along with our propensity for skin-deep loyalties. To be American is essentially to be from somewhere else. Robert Frost, the quintessential New England poet, was the California-born son of an Indiana "copperhead"—a Northerner sympathetic to the Confederate cause, notes historian George Pierson.[8] Our hero-rebels are mobile: the Indian nomad, roaming cowboy, fugitive slave, outlaw, and driver embracing the open road, as Sylvia Hilton and Cornelis van Minnen note in *Nation on the Move*.[9] The traits that foreigners both detest and adore in us— our informality, friendliness, instability, and restlessness—owe much to the idea of movement, according to Pierson. Spanish traveler Juan Bustamante y Campuzano wrote in 1885 of our "tireless activity."[10] Alexis de Tocqueville, just a month into his travels, concluded that Americans had a "restlessness of character." Later, he elaborated: "[I]in the United States, a man . . . plants a garden and [rents] it just as the trees are coming into bearing; he brings a field into tillage and leaves other men to gather the crops; he embraces a profession and gives it up . . . Death at length overtakes him, but it is before he is weary of his bootless chase of that complete felicity which forever escapes him."[11] Will we never tire of the hunt for something better? Is the world just now acknowledging that we all inhabit, as R. Buckminster Fuller gleefully wrote, "spaceship earth"?

Today, however, we move ourselves, not our houses. We are bodily "shunted from place to place, just like other goods," writes Nigel Thrift.[12] We have one of the lowest rates of residential mobility in the postwar era. Just 14 percent of Americans move annually, down from 17 percent a decade earlier. But the average number of miles that Americans drive annually has increased 80 percent in the past twenty years.[13] "Long weeks in a single community are unusual; a full day within a single neighborhood is becoming rare," writes Kenneth Gergen.[14] The local and global meld in a blur of trips—longer commutes, a vacation to Maui, business travel, a weekend flight to the Mall of America. Around the world, there are more than six hundred million international passenger arrivals each year, compared with twenty-five million in 1950. At any one time, three hundred thousand people are in flight above the United States. A half million new hotel rooms are built each year globally.[15] We can't all be what Virgin Atlantic, in a silky paean to its best customers, dubs the "fast-moving, culture-shaping jetrosexual, who dwells in multiple worlds and time zones," ever-hungry for the new and at home in the placeless metropolis of the airport. But that's no matter, because the *possibility* is there, and the yearning is deep. We have the tools of the trip—the cases of water in the DVD-equipped van, the gym clothes in the duffle bag, the charged gadgets, portable soundscapes, wheeled backpacks, satellite maps. There's a fine line between a car and an RV these days; for the first time, cars are coming equipped with closets. "Observe the contents of a stranger's briefcase," writes Cullen Murphy. "In all likelihood, the owner could be transported unexpectedly to Alma Ata or Baku and yet still retain full access to money, power and freshness of breath."[16] It's not so much moving for a job or a love or an ideal that stirs us anymore. It's moving for movement's sake. "Mobility climbs to the rank of the uppermost among coveted values," observes Bauman."[17] The bootless chase continues. What's around the corner?

Jaime Eshak swung onto a congested four-lane street in Boston, driving toward Harvard Square, steering with her knees.[18] She fished around in an insulated lunch pack on her lap and dug up a sandwich she had slapped together at home that morning: two slices of yellow American cheese on grocery-store whole wheat. She was also lugging a bottle of water, apple, yogurt, and two granola bars, one a forgotten leftover morning snack and another for the afternoon. As she drove and chatted, she absentmindedly took a bite out of her sandwich. This is Eshak's daily routine. Four days a week, she eats the same sandwich—cheese for now, following a long stint of peanut butter and jelly—then on Fridays, treats herself to a tuna sandwich and stick of string cheese from one of the seven 7-Eleven stores she supervises in the Boston area. Breakfast is grabbed and eaten as she readies her sixteen-month-old daughter for day care. This morning: two waffles and coffee. "Syrup, no. Plate, no," said Eshak, a tall, thin Midwest-transplant with high cheekbones and long black hair pulled back in a ponytail. "I was running around the kitchen with a waffle in my hand." For Ershak, eating rarely means taking a break or sitting down. Even dinners involve as little fuss as possible. Her husband, a car mechanic, will grill, or they'll stop at a sandwich shop. Time is scarce, leaving Eshak on a constant, guilt-ridden quest to find "decent" food that fits into their busy lives. She tracks the fiber in her bread, the ingredients in her daughter's organic baby food, the salt and sugar content of the many energy and granola bars her family consumes. "My hobby is nutrition," said Eshak, whose deep voice is equal parts brisk and friendly. But when there's a choice, portability wins. "I'd rather be eating and moving and doing something productive."

We were standing in a 7-Eleven store, her first stop of the day. Tucked on a busy urban street corner near City Hall, the store caters to office workers, the city's ubiquitous college crowd, and a sizable number of construction workers. As Eshak conferred with the manager about new signs for a sandwich case, customers trickled in, buying cereal and instant ramen noodles, newspapers and coffee. Almost everything edible in the store is made for eating on the run: single-serving cans of soup that can be heated in the store's microwave, hot dogs, tube-shaped pizzas, and sixteen-, thirty-two- and sixty-four-ounce Big Gulp sodas. Everywhere, portability is a selling point. A pack of two Milano cookies is labeled "On the Go!" A kit combining crackers, tuna fish, pickle relish, a mint, mixing spoon, napkin, and reduced-calorie mayonnaise is, naturally, a "Lunch to Go." In 2006, 1,347 products with "go" on the label debuted on the global market, a nearly 50 percent increase from the previous year.[19] Store manager Mekonnen Kebede, an Ethiopian who emigrated from Sudan at age sixteen, is amused by American eating habits.[20] In the Africa of his youth, offices closed at midday and people ritualistically went home or out for long lunches. He stood next to a six-foot section, center stage in the store, devoted to nutrition, energy, granola, meal replacement, and other bars—a market that's experienced double-digit annual sales growth nationwide since the late 1990s.[21] At this store alone, I counted ninety-five different flavors and types, from South Beach Diet to strawberry pomegranate. "Delicious Baked Goodness!" read one label. "We've taken the memories of home-baked banana nut loaf and turned them into a convenient on-the-go snack." (How much of my USDA daily recommended allowance of memories is in one serving?) Kebede walked over to a big refrigerator and yanked out a bottle of juice enriched with megadoses of protein and vitamins. "I have customers who drink this for lunch," he said, holding up the bottle. "That's lunch. That's it." He looked just as stunned as Eshak was when she heard that entire cities grind to a halt for lunch.

Food is fuel, and this is reflected in the look, feel, taste, and sensation of what we consume. Eating is not a moveable feast, nor even a picnic. Rather, foods are looking more and more like astronaut fare—concentrated, minimalized, single-serve. We buy crustless, frozen peanut butter and jelly sandwiches and dinner kits that come in their own bowls. (The side dish is going the way of the cloth napkin.) Just 47 percent of in-home meals include a "fresh" item, such as a vegetable, compared with 56 percent two decades ago.[22] Liquid meals, once the preserve of hospital patients, are in vogue. "Food is functional. It's a bar you eat and you've got your meal," says Kevin Elliott, vice president of merchandising at 7-Eleven, which is trying to transform itself from a mini-grocery where you pick up cigarettes and milk to a kind of gastronomic docking station.[23] At his Dallas office, he keeps a small blender, where he makes a yogurt and protein powder shake each day for lunch. "I'm good to go for the day," says Elliott, who also sees a big future in the stores for another of his new habits—doctoring bottles of water with protein and vitamin powders. "The consumer has a real need to start and to stop," he told me, stepping away from our telephone conversation for a moment to pay for a pizza delivery to his home. "They want to go, and they want to go fast, and when they want to shut down, they want to shut down quickly."

The imagery is unmistakable. As market researcher Clotaire Rapaille explains, "[T]he body is a machine and the job of food is to keep the machine running."[24] Pleasure in eating, writes Rapaille, "pales next to our need for movement. . . . We are a country on the go and we don't have time to linger over our food." Married mothers spend eight hours a week eating, an hour less than in 1965.[25] An American eating at McDonald's typically zips through a meal in fourteen minutes, while a Frenchman at the Golden Arches dallies an average twenty-two minutes.[26] In 1970, Americans spent $6 billion on fast food. By 2000, we were spending $110 billion annually—more than we spend on higher education, personal computers and software, new cars, movies, and most media.[27] And often, eating is a brief and solitary act. For all our earnest intentions, dinner regularly entails "one or both parents grabbing something on the way home from a late night at the office, one kid pouring a bowl of cereal for herself, and another heating something in the microwave on his way out the door," observes Rapaille, a French-born consultant with a doctorate in medical anthropology. "Americans say 'I'm full' at the end of a meal because . . . [their] mission has been to fill up their tanks; when they complete it, they announced that they've finished the task."[28] With a mischievous Gallic jab at his adopted country, Rapaille can't help but note the preponderance of food kingdoms sprouting within gas stations. (A gas station store used to be about seven hundred square feet. Now they are triple that size.)[29] "When you drive up to the pump and tell the attendant to fill up your tank," notes Rapaille, "it wouldn't be entirely inappropriate for him to ask, 'which one?'"

You can blame the fork for at least part of these predilections. Yes, the fork, the newcomer to the modern table, vilified upon its introduction to Europe in the late Middle Ages for being not only unmanly but, with its then-double tines, satanic. After a long, laborious journey to acceptance, the fork's star shone brightly for just a few centuries. Now, once again, "we dispense with most utensils," wistfully writes Barbara Bloemink in the catalog of a lavish cutlery exhibit at the Cooper-Hewitt National Museum of Design in New York. "We are again eating with our fingers."[30] But even if we set the table less and less, the fork has left a lasting mark on our eating habits. "The progress of the fork traces nothing less than the development of Western civilization," asserts food historian Darra Goldstein.[31] Certainly, the history of this upstart implement parallels the rise of our fervent individualism from premodern communality. Before we could invent single-serve nachos or soup cans for one, we had to get our hands out of the cooking pot. Before we could make mobility synonymous with success, we had to liberate ourselves from the need to stop and reunite to eat. To understand our place in the world, the fork is a surprisingly useful tool.

Back up a few centuries then, and take a place at a medieval table. Feasting "was a tumult of noise, color, and movement, a giddy mixture of ceremony and farce—the utmost refinement in art and manners side by side with animal appetites and crudities," writes geographer Yi-Fu Tuan.[32] Gnawed bones and dog excrement cluttered the floor, jugglers and servants bustled about the diners, and the table top was equally a free-for-all. "Cooks were singularly indifferent to the unique textures and flavors of the materials that went into a cauldron," writes Tuan, clearly aghast. A first course might consist of large platters of boiled and pickled boar, hulled wheat cooked with milk and venison, hart, swan, perch, and rabbit, lists Tuan, and be followed by a second course of similar meats and fishes piled high, pell-mell. In many senses, cuisine had little changed in the two thousand years since Roman times, when meals veered between elaborately presented whole animals and a "bewildering number of barely differentiated dishes served en masse." One rich concoction that the emperor Vitellius lovingly dedicated to Minerva, goddess of wisdom and skill in the arts of life, consisted of "pike liver, pheasant brains, peacock brains, flamingo tongues and lamprey roe." (Potent, although hardly portable.) Medieval diners may have sat up to eat while their Roman counterparts were prone, but their approaches to food were similarly indiscriminate.

Between the medieval and Victorian eras, an earthy, sensual communality gave way to a linear, scripted individualism. Sharing a scrap with the dog or a companionable tidbit with a dining partner, while hardly a new invention, was idealized in medieval times. Although knives had been fairly domesticated and spoons were in use since Roman times, even well into the fifteenth century well-bred people used their hands to dip into communal platters and bowls flowing with overcooked and liberally sauced meats, notes Goldstein.[33] Individual sets of utensils were not provided by hosts until the late 1600s, although some noble men and ladies traveled with their own. Elizabeth I owned forks but preferred her fingers, as did King Louis XIV. The long, slow infiltration of the fork into Western life mirrors Western society's increasing valuation of the self-conscious individual. Imported to Italy by a Byzantine princess, the fork garnered resistance both for its satanic look and for distancing diners from the godly gift of food, adds Goldstein. (The Catholic Church especially opposed its adoption.) Along with these sins, the fork gained a reputation as effeminate due to its initial use mainly for spearing "suckets," the sticky preserved fruits fashionable in court circles at the time. Long, lean, and sometimes topped by a spoon for catching juices, these early forks do look both dainty and illicit. Eventually, though, the controversial fork rose to prominence as standards of etiquette and hygiene developed that encouraged individual place settings and the separation of diners from each other and their food. "A civilized person was someone who consciously sought to put a distance between self and all that hinted at animality, natural functions, violence, messiness and indiscriminate mixing, whether of different kinds of food or of people," concludes Tuan, contrasting this with Chinese dining rituals that evolved to laud man's harmonious place in nature. "Rules of etiquette [in China] were intended to enjoin, not separate," notes Tuan.[34]

Now we've left the fork behind, the casualty of a time-pressed age. But while we again eat with our hands, we're rarely touching our sustenance. Wrappers, packaging, cans, straws, and the pace of life keep us from directly connecting with food until it's halfway down our gullets. And the food itself, of course, is many steps removed from the drippy, messy, and sometimes wholly recognizable fare that graced many a groaning table of the past. In the name of civilization, we've moved toward clean, processed, and *unobtrusive* foods. A quiet fill-up, that's what people tell Regina Lewis that they want. Nothing smelly, crackling, or noisy. We want food that takes a backseat to life—and we want it solo. In the 1970s, NASA hired the French-born industrial designer Raymond Loewy to help plan the living arrangements for the first space stations where astronauts would live for months at a time. To make the module homelike, Loewy suggested including a porthole, a space for private time, and a dinner table. American astronauts loved the view and the cubby but "hated" communal meals, writes environmental psychologist Jamie Horwitz.[35] They wanted to eat "on the run" and by themselves, while Europeans relished coming together to eat. One American space veteran suggested doing away with the table and installing automat-style food dispensers. That way, astronauts could tank up whenever they liked.

Are we neo-nomads then, powering through mobile days fueled by jet-set sustenance? Certainly, we are captivated by the allure of what T. E.

Lawrence called "that most deep and biting social discipline."[36] True nomadic pastoralism developed as an alternative, not a precedent, to settled, agrarian cultures. After small bands of hunter-gatherers began to settle, producing the first agrarian and later the first urban settlements in ancient Sumer and Mesopotamia, breakaway herders established a new cultural model based on man's age-old instinct for movement and newly found skills of husbandry. Nomads, in other words, consciously gave up an often impoverished life on the farm to inhabit the fringes of civilization, adopting and rejecting the ways of the settled to various degrees. They became the semi-tamed "other," appearing on the horizon to trade or plunder according to rhythms that mystified farmers and city people. The Chinese classified nomadic tribes on their borders as "raw" or "cooked," depending on the degree to which they had been influenced by Chinese civilization.[37] When one Hsiung-nu tribal ruler visited the Han Dynasty capital, he was housed in the zoological gardens.[38] Further off, Europeans were often more cowed than condescending. One medieval London monk, writing fearfully yet with begrudging respect for the Mongol tribes then ravaging Russia, described their horses as so dreadfully huge that they had to be mounted with stepladders and so voracious that they devoured branches and even whole trees.[39] (Actually, steppe ponies, while speedy, were on the small side.) By the late 1960s, when Europeans slept more soundly in their beds, art expert turned travel writer Bruce Chatwin wrote to his editor proposing a book on nomads. A chapter detailing civilized man's longings for nomadic life was to be called "Nostalgia for Paradise." Chatwin never wrote the book, but many of his works reflect his own brand of seminomadic existence.

Our infatuation with nomadic life has grown in direct proportion to the rise of highly complex, circumscribed, and crowded societies. Nomads are our alter ego, a personification of escape from the gray flannel suit and the tract home. Attila's use of a simple wooden cup despite great plundered wealth, the ingenious simplicity of the felt yurt, the Bedouin ability to thrive in a harsh desert climate—all are marks of the nomads' liberation from many fetters. British diplomat

John Ure writes of the long line of Anglo-American explorers, from T. E. Lawrence to Jane Digby, who tracked and studied nomads in past centuries. These adventurers, most of whom were semi-tamed rebels themselves, were obsessed with nomads' "freedom of movement, freedom from authority, freedom from the habitual anxieties of urban living, freedom from the constraints of organized agriculture, freedom from any convention but their own."[40] Freya Stark, a flamboyant self-taught Arabist, came to prominence in the 1930s when she tracked down an ancient, secretive Muslim sect that committed political and religious murders from their bases in Syria's Valley of the Assassins. Bruce Chatwin, who had a degree in archaeology, set off on the first of his travels after temporarily losing most of his sight. "People would compliment me on my 'eye,' and my eyes, in rebellion, gave out," writes Chatwin.[41] The doctor could find nothing wrong, so he made a quaint suggestion. "Perhaps I should try some long horizons? Africa, perhaps?" And so began Chatwin's escape from the daily grind to follow the nomad's tantalizing trail.

Perhaps the most hopelessly romantic among such travelers is Isabelle Eberhardt, a hashish-smoking, free-loving, cross-dressing Russian expatriate who roamed Algeria at the turn of the twentieth century, filling her diaries with astute observations of a region that few Europeans had seen. Eberhardt, who drowned at age twenty-seven in a flash flood in the Sahara's Atlas mountains in 1904, alternatively called herself a "rootless wanderer" and a nomad and accurately saw a life of movement as both her destiny and her doom. "To be free and without ties, a nomad camped in life's great desert, where I shall never be anything but an outsider, a stranger and an intruder, such is the only form of bliss, however bitter, [destiny] will ever grant me," writes Eberhardt. "A nomad I was even when I was very small and would stare at the road, that spellbinding white road headed straight for the unknown and I shall stay a nomad all my life, in love with changing horizons. . . ."[42] Her idealistic, counterculture outlook would not be out of place today, in an era when we call ourselves "cyber-" or "digital" nomads, imbuing our virtual journeys with the dignity and weight of this ancient way of life. "The potential for nomadism has come about as the result of a series of liberations," from telecommunications monopolies, geographic ties, even repressive regimes, claim David Manners and Tsugio Makimoto in *Digital Nomad*.[43]

Pastoral nomads, of course, are not "rootless wanderers," free to go where they please. Their harsh, often chosen way of life is defined by movement for survival's sake, not exploration or a quest to live "without ties." Most pastoral nomads, diverse as they are, follow seasonal pathways, moving back and forth to graze their flocks in the best pastures. "Every year when summer came, they would set off back to the mountains," writes historian Fernand Braudel of the nineteenth-century shepherd tribes of Greece. "Behind the long procession of sheep, whose speed dictated that of the rest, came the convoy of horses, up to a thousand at a time, all laden with household goods and camping materials, tents—and young children sleeping in baskets. There were even priests accompanying their flocks."[44] Such movements were collective, as well as highly organized. Owen Lattimore, an American raised in China by missionaries, points out that steppe nomads of Central Asia had to cope with the differing grazing needs and travel patterns of yaks, camels, horses, mules, and sheep. Families depended closely upon each other, and all relied on a deep knowledge of the land. Unlike the Chinese, who rode single file for long distances, steppe nomads rode alongside each other or spread out in a wide horizontal line, so that they could disperse easily if attacked, their horses could graze, and also so that riders could closely observe the landscape, notes Lattimore, who traveled in Mongolia for four decades.[45] "If Lattimore—exhausted from the day's march—fell in behind the leading camel and became silent, he would be chided with behaving like a Chinaman," writes Ure.[46]

Skimming on the margins of our old settled ways of life, we are like pastoral nomads. We, too, go and return, rising before dawn to endure long, congested journeys that we retrace at day's end. The number of commuters traveling more than sixty minutes each way to work grew nearly 50 percent to 8 percent between 1990 and 2000.[47]

We shuttle between increasingly distant first and second homes, each little used. But we are also insatiable travelers on a canvas writ large, and so living a mobility that is looser, transient, and akin perhaps to a primeval form of nomadism, that of the hunter-gatherer. Hunter-gatherers are defined by *focus*, not the *boundary*, writes Peter Wilson in his *The Domestication of the Human Species*. "Without boundaries and without the *concept* of the permanent boundary, people are not conceptually locked into their relationships or surroundings," according to Wilson, an Australian anthropologist who argues that architecture changed the nature of attention in important ways. "Nomads focus on the landscape and its features; sites, tracks, water holes, lairs, sanctuaries, birthplaces, landmarks." (Today, read: malls, highways, drive-thru restaurants, offices, billboards, both earthly and virtual.) "Nomads focus on one another in the same way, keeping those in clearest focus who live in intimate proximity and for whom, at the time, they have intense feelings."[48] Relationships are fluid, a preface to ours in the boundaryless postmodern world, and self-sufficiency is highly prized, again as it is in our own day. This is not a formless, drifting life, yet it is one in which routines and traditions are far less "marked, concentrated, confined or bounded" than in sedentary society. A society defined by movement, not settlement, does not prize permanence. "Nomads have neither past nor future," writes Claire Parnet. "They have only becomings, woman-becoming, animal-becoming, horse-becoming; their extraordinary animalist art. Nomads have no history, they only have geography."[49] Wilson notes that archaeologists studying hunter-gatherers "find little, if any, remnants of a camp a week or so after its abandonment. But archeologists of prehistoric village societies can reconstruct entire settlements thousands of years after they have been deserted."

Surely, our brand of nomadism is rooted both in this idea of the subversive "otherness" and in the notion of a protean existence. From this "becoming" arises what Pico Iyer calls the Global Soul: someone who grew up in multiple cultures or who lives and works globally, nonchalantly jetting from "tropic to snowstorm in three hours." Iyer is a Global Soul, having been raised in California by Indian immigrants who sent him to English boarding schools from age nine. The Global Soul, writes Iyer, cultivates a "porous self of self" and a mutable sense of home. "We live in the uncertainties we carry around with us," writes Iyer.[50] As the world grows "flat" in Thomas L. Friedman's words, aren't we becoming in many ways a culture of Global Souls? There is a sense of wonder and endless possibility to this polyglot life, shaded by the uncertainty that is the traveler's constant burden. The Global Soul is proud of her ability to feel at home anywhere and yet dreads turning a corner and finding her fragile sense of equanimity torn from her.

After our Dunkin' Donuts lunch, Regina Lewis recalled using a mapping Web site to find a running route while on a weekend trip to Portland. "Here I am in Maine and I know I'm going exactly eight miles and it maps my run and I feel completely comfortable and I know exactly where I'm going," she gushes. (Freedom of choice, says sociologist Zygmunt Bauman, now has a conspicuous spatial dimension.)[51] Lewis and I began discussing the perpetual traveler's need to equip herself, and Lewis recounted a talk that she had recently attended on how mobile professionals create a sense of home on the go. Suddenly, she leapt off her cafeteria stool, miming how she—like other global nomads—hides hotel literature in drawers and places her own book and clock on the nightstand when she checks into her room. "That clock appears to be fantastically important to all of us," said Lewis. "For one thing, it's our link to time and just how our life is passing us by and for whatever reason, when we travel a lot between different places and hotel rooms, we want not only a clock, but we want our personal clock, which makes us even more tied to our own sense of life." (*"How our life is passing us by"?*) More than a talisman, the clock anchors us in a universe of, as Bauman says, "moving sands."

Global nomads' biggest fear is waking up in a hotel room and having no clue where they are, says Fleura Bardhi, the marketing expert whose presentation Lewis attended. For years, Bardhi has been studying aid workers, environmentalists, bankers, and others who travel at least 60 percent of their time, experiencing in full the mobile lives that we're adopting bit by bit. Some of these travelers have little connection to their current residence but idealize their childhood home or homeland as a place to eventually return to in old age. Others are expatriates who temporarily anchor themselves abroad, keeping a "museum of the self" back in their native country. A final segment is entirely rootless, except for one crucial possession, such as a baseball cap, pocket knife, or photo, that is carried at all times, an "anchor of the self," says Bardhi.[52] These are worldly people, with temporary landings, portable ideas of home, and careful rituals constructed to dispel a sense of drift. And Bardhi understands them well, for she is one of them, having migrated from her native Albania to Norway, then Nebraska, and now Boston for her studies and work. Upon moving to Boston, she chose to live in a venerable Irish, now gentrified neighborhood where she could feel part of the city, yet anonymous. "It's important for me to establish a quick sense of home. You can leave the next day and not miss your neighbors," says Bardhi. "I can't allow myself to fully be comfortable, feel at home and identify with a place." In his essay collection *Anatomy of Restlessness*, Bruce Chatwin recalls meeting a British typewriter salesman who had to visit all the countries of Africa every three months, with a stop in London at the close of each orbit. This man lived out of a suitcase, but in his company's safe in London, he kept a small black tin box, filled with his "things:" a teddy bear, photo of his deceased father, childhood swimming trophy. Every time he came back to London, he threw one object out and added another. "He is the only man I ever met who solved the tricky equation between things and freedom," writes Chatwin wistfully. "The box was the hub of his migration orbit, the territorial fixed point at which he could renew his identity. And without it he would have been literally deranged."[53] Portable clocks and little black boxes. How little it takes to ground us. And yet, how little grounded we are. Movement has its perils, and foremost among them is the fact that in moving too far or too fast or too often, we literally risk losing ourselves.

We struggle now to update our ideas of home and place for a new unbounded era, naturally wishing to shed outdated notions of stasis and rootedness. We relate less and less to place in these ways, perhaps partly because of the well-tracked rise of homogenous, transient place-less places—from malls to chain stores. And yet an evolving and per-haps increasingly elusive sense of place does not make place any less important. Humans will always need space as a canvas for movement and venture, asserts Tuan, and place as a "calm center of established values" or, more simply, a realm of "pause."[54] Traditional nomads have always been bonded to landscape and ritual, even while spurning settlement. Place shows us where we belong and is essentially about attachment and involvement. The making of place demands time. "If we see the world as process, constantly changing, we should not be able to develop any sense of place," says Tuan, relating an illustrative story of how Hamlet's legendary association with Kronberg Castle in Denmark transforms our view of this building. "Isn't it strange how this castle changes as soon as one imagines that Hamlet lived here?" the physicist Niels Bohr said to his colleague Werner Heisenberg while they were visiting.

> As scientists, we believe that a castle consists only of stones, and admire the way the architect put them together. . . . None of this should be changed by the fact that Hamlet lived here, and yet it is changed completely. Suddenly the walls and ramparts speak a dif-ferent language. The courtyard becomes an entire world, a dark corner reminds us of the darkness in the human soul, we hear Hamlet's "To be or not to be." Yet all we really know about Hamlet is that his name appears in a thirteenth-century chronicle. No one can prove that he really lived, let alone that he lived here. But everyone knows the questions Shakespeare had him ask, the human depth he was made to reveal, and so he, too, had to be found on a place on earth, here in Kronberg. And once we know that, Kronberg becomes quite a different castle for us.[55]

Places link us to history, literature, the murky depths of the human soul, and to others. A shared thread in the tales of Bardhi's travelers is their propensity to leave broken relationships in their wake. Place is, in Tuan's words, a "field of care."[56]

When asked why he became a geographer, Tuan gives various answers. At parties, he often simply says that as the son of a Chinese Nationalist scholar-diplomat, growing up in the 1930s and 1940s, he moved often. On more serious occasions, he explains that he always had an "inordinate fear" of losing his way. "To be lost is to be paralyzed, there being no reason to move one way rather than another," he writes in his autobiography, *Who Am I?* "Geographers, I assumed, always knew where they were. They always had a map somewhere—either in their backpack or in their head." His candor is poignant, and telling. The Oxford-trained Tuan went into his chosen field at a time when geographers tended to be rugged Indiana Jones types, adept at slogging over mountain and desert to decipher and map the physical realities of earth. Tuan, a slightly built, brilliant thinker, instead broke ground, so to speak, by exploring the landscape of human experience. He revolutionized his field by studying the links between our inner and outer worlds, from the way a mother acts as a primal "place" for her baby to a cathedral's embrace of all our senses. In particular, he led geographers beyond a narrow focus on the characteristics of places, to a consideration of the *meaning* of place. Rarely, Tuan offers his deepest reason for becoming a geographer: he wanted to know "what we are doing here, what we want out of life." Looking back over his life in his seventies, he observes that he has lived perpetually on the intellectual edge, probing for new insights to unanswerable questions, probing in essence to find himself. He is content with his brand of "deviance," which has had a powerful effect on postmodern geography. And yet Tuan has never felt at home in the world or found something akin to "permanence." His life is striking both for the courage of his intellectual explorations and the constancy of his loneliness. "Among the major causes of a weakening sense of self are social and geographical mobility and rapid technological change," he writes in the opening pages of his autobiography. "We are, as pundits say, in the midst of an identity crisis."[57] This, in essence, is the explorer's tradeoff, as Tuan, Stark, Chatwin, Iyer, and others before them knew well, and we are all discovering.

We are becoming a nation of the untethered. Is this why we live in what cartographer Denis Wood calls the "Age of Maps"?[58] Are we turning to maps to find our way and perhaps ourselves? Certainly, maps have always been a tool for ordering life, hence an instrument historically used by the powerful to filter and shape perception. We rely on maps to reassure us of the perimeters of the known world and to shed light on the unknown. Just as maps reflected Europe's dread and delight of the New World, they are a singularly perfect medium today for attempting to grasp the amorphous new frontier of cyber-space, and especially for visualizing the interplay between the virtual and the physical. Computerized "geographic information system" tools layer information—from infectious disease hotpoints to electricity grids—over maps. Conversely, "geocoding" pastes geographic coordinates, such as longitude and latitude, over online text or even sound files. How far we've come from a century ago, when the 1910 *Encyclopedia Britannica* defined a map as a "representation, on a plane and on a reduced scale, of part of the whole of the earth's surface."[59] The map has wriggled off-paper and gone animate and protean as never before and so is both an echo of and a siren call to our neonomadic journeys. (An explorer always has a map in her head or backpack.) No wonder, then, we seize them delightedly; 99.9 percent of the maps ever made, estimates Wood, were created in the past century. While big companies, from Google to Esri—the Rand McNally of GIS—largely control the new multibillion-dollar business of high-tech mapping, we are appropriating simple Internet-based cartography tools and their veneer of authority with glee. Not long ago, when cartographer Margaret Pearce told people about her work, they'd often profess a love of maps. In recent years, people say, "'I make maps, too,'" relates Pearce, a professor at Ohio University. "Now people see themselves as mapmakers."[60]

A plain-spoken Midwesterner with gentle eyes and long, straight oak-brown hair, Pearce seemed a bit bewildered to be sitting in the Lucky Cat Lounge in the trendy Williamsburg section of Brooklyn one rainy Saturday afternoon. The dark, cramped nightclub was one of the main venues for Conflux, an annual, four-day arts festival devoted to "psycho-geography"—the physical and psychological exploration of the urban landscape. As I sipped coffee with Pearce, hordes of artists, geeks, hangers-on, and students drifted in and out of the Lucky Cat en route to lectures, performances, walking tours, and street games. There was an artist making a show out of visiting a park for what he said was the last time; a waterfront olfactory peregrination; the slow building of a wobbly-looking desk on the street; a walking tour using a "mash-up," or subversive retooling, of a map of Baghdad. The festival is hip and gaining a cult popularity reminiscent of an early Burning Man. It's far removed from drive-thru pizzas, endless freeways, and placeless places. And yet from Conflux's energetic swirl of performances arise intriguing truths about our shifting relationship to the earth in an age of distraction. For mapmaking is essentially a symbolic expression of our connection to the physical world. Mapmaking, at its core, is a form of attention. Can digital maps counterbalance our relentless mobility and reconnect us to the earth and to our bodily selves? Can we create virtual songlines for our time?

Nomads traverse the earth lightly. Attila the Hun's hillside capital in modern-day Hungary featured a log palace and a city of wooden huts but only one building of stone: the bathhouse. The Mongol royal capital of Kharakhorum was a tent city, with few permanent build-ings.[61] Yet many nomadic cultures, the Australian Aborigines most famously, construct elaborate mythical topographical maps that anchor them firmly to the landscape. Aboriginal songlines are deeply individual, dizzyingly complex grids of ancestral tracks and sacred sites that are of inestimable importance in establishing individual and collective identity. They are, in a sense, a nomad's invisible architec-ture of place, an unbroken link between a people and their landscape and history. In nomad culture, "a construction is put on the landscape rather than the landscape undergoing a reconstruction, as is the case among sedentary peoples, who impose houses, villages and gardens on the landscape," notes Peter Wilson.[62] Carrying on this tradition, Pearce seeks to make maps that tell stories as a means of forging intimate connections to place. "Cartography invariably gets used for the depiction of space, not place. But we *can* represent sense of place in geography," she told festival-goers, showing them a colorful, six-foot map she created that breaks just about every rule of traditional cartography. The map depicts an eighteenth-century fur trader's first canoe voyage from Montreal into the Canadian interior. Pearce, however, shows much more than the lay of the land along his route. She boxed off each day's progress and put excerpts from his diary directly onto the map in various hues and fonts to show weather and emotion. She is deliberately trying to depict the qualities of place that hold the narrative of a journey. She's trying to create songlines for our time. Is this what we're doing when we virtually tag nightclubs or leave Web-based poems and photographs on street corners?

On the afternoon before I heard Pearce speak, I joined the Baghdad mash-up tour "You Are Not Here." A dozen people set out from an art gallery with a student from New York University's Interactive Telecommunications Program who helped created the event. We carried a double-sided map: on one side was Baghdad, on the other New York City, mapped in a one-to-one ratio. If you held the map up to a light, you could see how each stop on the walking tour corresponded to a destination in the Iraqi capital. Mushon Zer-Aviv explained to me that the mash-up is intended to "add another layer" to people's media-dependent experience of Iraq, and counter Western tourists' consumptionlike "ownership" of places they visit.[63] "They are not interested in the true story," said Zer-Aviv, an Israeli. Once we got started, it took quite a while to find the first stop, since Zer-Aviv stuck closely to his scripted role of being just another traveler. "It's my first time in Baghdad as well," he said as we traipsed through the gentrifying neighborhood. "Who are we following? I don't speak the language." Finally, we found a "You Are Not Here" sticker with a phone number posted on a broken lamppost in a desolate waterfront area. We called in to hear a prerecorded commentary on the famous toppling of a statue of Saddam Hussein on this "spot." "Where should we go now?" asked Zer-Aviv. "Does anybody know what we're supposed to look for now?" Leaving the tour, I headed back to the subway, past empty lots and dark warehouses, to a street where stout Polish women with lined faces ducked into kielbasa-scented butcher shops to do their Friday shopping. At least for me, the tour lived up to its name. I didn't learn about Baghdad, nor did I truly focus on the sights and sounds of this vibrant Brooklyn neighborhood. I just drifted through, distractedly. I was not really here.

The Lucky Cat was jammed when Denis Wood took the stage Saturday night, and a worshipful crowd, decades his junior, spilled across the floor and past the hopping bar. Wood is an irascible cartographer and professor who's spent time in prison and spars intellectually as if he's in a knife fight. He earned his master's and doctorate in the early 1970s from Clark University's world-famous cartography program, and has kept his rebellious radical outlook intact. He's a Grand Old Rebel now, and the festival-goers at Conflux sat enrapt as Woods read a rather dry paper about one of the fathers of psycho-geography, Guy Debord. The French filmmaker and poet espoused fighting capitalism and modern alienation by breaking free from societal roles and infusing everyday life with creativity. To see society anew, Debord and his followers, the Situationists, went on urban "derives" or "observational drifts," a technique that is cousin to Wood's own famous, perspective-shaking maps of pumpkins on front stoops, puddles of light from streetlamps, sewer pipes, and power lines in his Raleigh, North Carolina, neighborhood. "Mapping is a conscious-raising, perception-refining, radicalizing activity," says Woods, who was one of the first to rattle the staid cartography world with the idea that maps should tell us about the experience of place. At the end of his talk, though, he chastised the would-be psycho-geographers for losing their way. Clicking, Googling, and manipulating maps isn't the same as sensing the world anew through mapmaking, he growled. Everybody loves to make maps nowadays "as long as they can find the data on the web. Behind the map is the actual collecting of the data—the walking out on the street and looking and smelling and hearing and making those annotations, and what it means. You've got to open your eyes."

Are we creating songlines as we zoom astronautlike around the satellite world of Google Earth or sprinkle virtual Post-it notes on street corners for the cognoscenti to encounter? Or are we using the earth as our stage set, making cut-and-paste constructions upon the landscape? Surely, our maps echo our veneration of exploration, our facility with place shifting, our enchantment with posing new questions and storytelling. But a society that leaves no room for attachment cannot make songlines. We are not using maps to ground ourselves but rather to enable ourselves to keep moving on. On a walk through Williamsburg, Wood told me that he fears Americans are losing a sense of place. "I don't find it in the subdivisions, in the cul de sacs," he said. "I don't find it in the strip malls, on the highways, certainly not on the Interstates and certainly not in the office parks."[64] We stopped for a moment, and I asked if this neighborhood, with its quirky blend of old and new, would be easy to map. "No," he said. "The range of architectural detail is extraordinary, everything's idiosyncratic, everything's different—these are serious mapping challenges." Wood notices everything. He drew my eye to trash cans, a dusty construction project, ornate cornices, bars, and shops. "It would be a hard place to map," he concluded, taking a last look around before heading back to the Lucky Cat. For a rare moment, he was tranquil.

Detachment is the cost of our wondrous, liberating mobility, the price we pay for living untethered. Connection to place is replaced by inhabitation of space, bringing to near-culmination what Rosalind Williams describes as "humanity's decision to *unbind* itself from the soil."[65] The tempo of travel blurs the landscape, and our vehicles increasingly enfold us in a bubble of remove. "The sights, sounds, tastes, temperatures, and smells of the city and countryside are reduced to a two-dimensional view through the car windscreen, something prefigured by the railway journeys of the nineteenth century," notes John Urry, echoing Nietzsche.[66] The portable soundscape of the

Walkman-turned-iPod puts much of life at arm's length, creating a "fragile world of certainty within a contingent world," writes Michael Bull in *Sounding Out the City*.[67] We are distancing ourselves from knowledge of our own bodily selves and our earth, as Bill McKibben observes. We live in an era when "vital knowledge that humans have always possessed about who we are and where we live seems beyond our reach," he writes. "An unenlightenment."[68] Pause is increasingly absent in a temporal sense, too. A culture of constant movement, in part fueled by a love of instant gratification, cannot bear the mystery and unpredictability inherent in the idea of pause. "For the sake of speed, in the interest of not wasting time, we sacrifice the sensuous richness of the not-yet," writes Noelle Oxenhandler in her essay "The Lost While."[69] We live in a culture of "becoming" but never arriving.

We float, anesthetized even to our own bodily selves. We've moved from a premodern world of competing and conflicting sights, odors, sounds, and tactile experiences to an era of separate sensory experiences dominated by our most intellectual and distanced sense: vision, argues Tuan. As a result, "the perceived world expands," while "its inchoate richness declines."[70] From the cacophonous, slippery, odiferous, pungent eating of centuries ago, we've moved toward a gastronomy based less on sensory experience. We often literally don't *notice* our food, although it is sometimes our last connection to the soil. When a recent governmental survey of daily activities neglected to take multitasking into account, the results showed that nearly 10 percent of Americans reported no daily time spent eating—because they were eating only as a secondary activity. And we like it that way: 32 percent of women and 22 percent of men would prefer to consume their food in pill form.[71] Many young aren't accruing "taste memories"—the remembrance of a just-picked tomato, a juicy summer peach, food researcher Carol Devine laments.[72] Surely, they'll have memories of eating, although not exactly of real food, suggests Michael Pollan in *The Omnivore's Dilemma*. With its enticing jolt of fat and salt, a chicken nugget, for example, is a "future vehicle of nostalgia—a madeleine in the making" for children of today and yet, like all fast food "the more you concentrate on how it tastes, the less like anything it tastes," writes Pollan.[73]

All this is no surprise to Brian Wansink, a mischievous, towheaded Cornell professor who studies the psychology of indulgence. His myriad experiments show that, as people pay less attention to what and how they eat, they are increasingly prey to marketers of abundance. In one experiment, Wansink and his students rigged a soup bowl so it would never empty. Unwitting subjects just kept eating until the experiment was halted. At the movies, while watching television, at restaurants and parties, people will lap up whatever Wansink serves, no matter how copious, stale, or bland. "I'm really interested in low-involvement decision-making," says Wansink, who is both ruthlessly practical and boyishly charming. "The real problem is that there are a lot of things we have in society that encourage detachment."[74] Wansink himself is an omnivore's omnivore, gulping coffee one minute, popping a handful of M&Ms or a stick of green apple gum into his mouth the next. A perpetual-motion workaholic, he nevertheless sits down to dinner nightly with his Taiwanese wife, toddler, and, on the day I shadowed him, his elderly parents, visiting from Iowa. Growing up in a close-knit farm family, Wansink has always been fascinated by the role that food can play in creating a cohesive society and refuses to take part in the ideological battles pitting food snobs against mainstream America. The answer to better eating, he believes, lies in getting people to open their eyes and awaken their senses to the grotesque portions scarfed down distractedly and the tastelessness of processed food. Once early in his career, he got an obese man to break his addiction to drive-thru eating by persuading him to stop by the side of the road to eat. The man realized that the food simply tasted awful.

Can we stop now? Can we pull to the side of the road and look hard at the encroachment of the placeless places, or are we too enamored of our untrammeled solo journeys, the possibility of yet another choice around the corner? It's not too late. If we pause, we can begin to see that the land of distraction is a topography of diffusion, fragmentation, and detachment. We are pulled away to an alluring virtual realm and are growing sated by spectral glimpses of one another. We endeavor to hustle past the limitations of the clock via a split-focused life, and we lose our anchoring and hence our sense of self in a blurred life broken from place. Are we evolving toward a new space-age existence, shedding our biological limitations? Or are we too numb to see a coming decline? "What is needed is a new emphasis on the real, palpable, enfleshed materiality of our surroundings," asserts British critic Marina Warner.[75] What is needed is a way to counter our detachment from the world, recover the strength of deep connections, salvage wholeness in perception and thinking. In this land of distraction, we begin to rely on fragments, snippets, and push-button answers, and that is not a step forward. That is the beginning of a cultural decline.

At dinner, Wansink ate slowly, taking several small helpings of his wife's sautéed chicken, tomatoes, and fresh basil, and his mother's chopped bacon and broccoli salad. He slipped his one-year-old a chickpea, prompting a scolding from his wife about choking. The meal was leisurely, punctuated with tidbits from their day and stories from Wansink's childhood. Then the table was cleared, and his wife brought out Chinese sweets to celebrate the October harvest festival and the legend of the lady banished to live on the moon for her sins. There were tiny bits of dried mango, coffee-flavored dried plums, and little almondy cakes shaped like a full moon. Wansink put a tiny morsel of a cake on his daughter's highchair tray, and she eagerly popped it into her mouth, making such a joyful gurgle that everyone at the table couldn't help but laugh. The unhurried moment hung in the air, round and sweet as a moon cake.

Part II
DEEPENING TWILIGHT
Pursuing the Narrowing Path

Chapter Five

FOCUS

Invisible Tethers: The Delicate Art of Surveillance-Based Love

J eremy Bentham is still keeping watch. The political philosopher died at age eighty-four in 1832, but his preserved skeleton, topped by a wax head and dressed in his own frock coat, ruffled shirt, and wide-brimmed hat, resides in perpetuity in the main building of University College London. (His real head, mangled during a failed mummification, is stored nearby in an oak box. For a century it lay on a tray at the philosopher's feet, until it proved subject to one too many college pranks.) Per his stipulation, Bentham sits in his own straight-back chair, his favorite walking stick resting on his knees, gazing back at gawkers from a glass-fronted wooden box, an eerie testimony to his faith in the power of permanent visibility in all realms. He is now an "auto-icon," the term he coined for a "man who is his own image, preserved for the benefit of posterity."[1] And not surprisingly, the social critic and legal reformer had bigger plans for his auto-icon than resting quietly in a case like a preserved butterfly. Along with leaving detailed instructions for his corpse's public dissection by his personal physician and for the subsequent preservation of his body, Bentham decreed in his will that his auto-icon should be brought to meetings of his friends and followers "to be stationed in such part of the room as to the assembled company shall seem meet." In 2004, the auto-icon attended a university retirement party. Six years earlier on the two hundred and fiftieth anniversary of his birthday, the preserved Bentham was beamed via videoconferencing to a Texas symposium, where scholars addressed their remarks directly to him.[2] At day's end, participants sliced into a celebratory cake shaped like the Panopticon, the model for prisons, schools, hospitals, and other institutions that is one of Bentham's lasting legacies to our world. For as the creator of the "all-seeing place," Bentham is a father of our surveillance society. He first teased out what's now an established modern tenet: keeping people under watch is an ingenious way to regulate them. The gaze is mightier than the whip. Of course, Bentham assumed *he'd* be doing the watching, even after death.

The Panopticon—a circular building made up of cells arranged around a tall central observation tower—is a remarkably simple yet powerful notion. While working as an estate manager for a Russian prince in the late 1700s, Bentham's younger brother Samuel, a brilliant inventor and engineer, concocted the system as a way for one man to simultaneously supervise large numbers of craftsmen and factory workers. The occupants of the cells do not know when they are being watched from above, so the mere implication of surveillance becomes a potent form of control. During a visit to Russia, Jeremy Bentham eagerly set about adapting his brother's model for possible use in a range of institutions, but especially for prisons. Upon his return to England, he spent two decades promoting the idea of a Panoptic penitentiary, which he believed to be not only highly secure but rehabilitative. He bought and set aside a choice piece of Thameside land on what is now the site of the Tate Gallery and worked with Samuel to convert part of their house into a workshop for models of prison machinery, such as a huge treadmill for prisoners. No detail was too small for Bentham to hammer out, from the choice of construction materials—fireproof brick or iron—to the best prison trades for inmates.[3] To his bitter disappointment, no Panopticon was built in his day, and yet his auto-icon has survived to witness the rise of the invisible gaze as a central notion underpinning the management of society —and increasingly, our family life. Think surveillance cameras, software monitoring, GPS tracking devices, "black box" vehicle monitors, Breathalyzers, drug tests, implanted data chips in airports, parks, offices, hospitals, factories—and now at home. Surveillance is no longer simply a given in public; it's becoming the norm in "private" life. Spouses monitor spouses, the jilted track ex-lovers, children watch peers, and most of all, parents bathe children in a climate of the gaze. The government is watching, yes, but more than ever we are watching each other. The reclusive, obsessive Bentham was a visionary, indeed.

In a time of distraction, the gaze is the tether securing us to those we love. The word *surveillance* comes from the French to "watch over," meaning both to care for and to monitor. Through surveillance, we seem to gain a measure of security—an equally double-sided concept—in a world of insubstantiality and change. "The rise of surveillance societies has everything to do with disappearing bodies," notes sociologist David Lyon.[4] Watching and tracking and monitoring provides comforting evidence—the snapshot, the print-out, the fix on a map—of "presence" in a virtual, mobile, split-focus world. The invisible gaze makes the absent other somehow "visible," and hence safe. Where was he? What did she say? How long did it take? Most importantly, is he being good? Surveillance gives us answers in a world of unanswerables. And yet in turning to surveillance as a way of holding the family together, we pay a price that Bentham could not have predicted or understood. For surveillance can't cohabitate with trust, that slow-to-bud, immeasurable essence of close relations that thrives only outside the panoptic gaze. By choosing surveillance-based attention, we are ushering in an age of mistrust. This is the first collective loss we will suffer by cultivating a culture of distraction. Bentham's emblem for the Panopticon was an eye encircled by the words "Mercy, Justice, Vigilance."[5] But his Inspection House was built on a foundation of unforgiving control and an absence of trust. "I will watch until I observe a transgression," wrote Bentham. "I will minute it down. I will wait for another; I will note that down too. I will lie by for a whole day. . . . Learn from this, all of you, that in this house transgression can never be safe."[6]

> In the area where we've aimed the camera, we can see her lying in her bed, we can see if she gets out of bed. We don't rush in there if she gets up. Just a door opening and a little creak on the hinge is enough to scare her.

Jim,* a Midwestern technology executive, is describing the camera he set up in his two-year-old daughter's bedroom on the day she moved from a crib into a "big girl" bed. It all happened the second time that Zoe climbed out of her crib. Jim and his wife took action, assembling the bed they'd bought for her, then installing the infrared camera they'd kept on hand for that very moment. "Being first time parents, we were very interested in what was going to happen when we closed the door. She can be pretty resourceful. Was she going to go to an outlet, even though we have covers? Was she going to jump up and go through her books?" he recalls wondering. "It was really that first experience we were most afraid about. That's where the anxiety was." On the first night that they tucked Zoe into her new bed and shut the door to her room, her parents sat before the den television, watching a grainy, black-and-white, real-time feed of Zoe on the channel they now call "ZTV." As they watched, she began to climb out of her bed.[7]

Our world is no more risky than in the past. Each era has its hazards. Which is more unnervingly perilous, the Black Death or epidemic cancer, the threat from ruthless highwaymen or terrorists of our time? Yet the measurement and management of risk is more central to our culture than ever before, as the unavoidable cost of navigating a world that we feel that we can actively shape, says sociologist Anthony Giddens. Most of us no longer simply trust ourselves to fate. Like Regina Lewis, we are explorers. "The idea that you can control hazard and therefore can be insured against it—the rise of the notion

---

*Names of this family have been changed.

of humanly engineered safety—is part and parcel of Enlightened thought," says Giddens.[8] Managing risk, in short, is a way of organizing the future, and surveillance is a natural offshoot of our now-obsessive efforts to control what's coming around the corner. We make our children wear tricycle helmets, flotation devices, and seatbelts. We keep their fingers out of the cookie dough, rear them on well-done burgers, warn them of kidnappers, and relentlessly watch them in pursuit of that ever-elusive goal—"peace of mind." Surveillance is a weapon of choice in the war against risk, a kind of invisibility cloak of protection, with safety and control the inseparable, hoped-for outcomes. Without control, safety doesn't seem possible. Amid risk, control is lost. This is why it's natural to think that keeping children safe today means controlling them. Yet in the panoptic society, the line between vulnerability and transgression fades. The innocent get watched for their own protection, but the watching invariably strips them of their innocence. Merely by stepping out of the "safe zone," the protected becomes the transgressor. "We've put kids under what I would call house arrest," says Steven Mintz, a historian of childhood.[9] Who is protected? Who is caged?

Dan Pope, a television meteorologist and father of three boys from Salt Lake City, set up software monitoring programs at home after discovering that his youngest son, age twelve, was unknowingly involved in an inappropriate online relationship with what appeared to be an older man posing as a young girl. Noticing the boy online at odd hours, Pope became suspicious. "I started to look over his shoulder," says Pope, who also enlisted his twenty-year-old eldest son to retrieve his younger brother's e-mails and instant messages. "I was naive."[10] Pope never found out who the stranger was, but he was sufficiently alarmed at the time to take away his boy's computer privileges for nearly a year before discovering a software program that allows him to monitor his son's computer use from a distance. Some Web sites are blocked outright, while others trigger a warning e-mail to Pope. He then checks out the site and, if it involves heavy metal music or other forbidden realms, tells his wife to oust their son from the computer for the rest of the day. "We're not trying to control him, to take his freedom away, or rather, we've tried to do it in a loving way," says Pope, whose son blew out his computer trying to disable the software. "It's really done with thought and with care and love."

Along with the slippery perimeters of cyberspace, we micromanage the boundaries of our physical realms, unwittingly turning ourselves into intruders in our own safe zones, as Steven Flusty writes in his essay "Building Paranoia."[11] Until the late 1980s, Flusty could enter the suburban Los Angeles house of his childhood by simply crossing the yard and turning a key. Now, one foot on the front walkway, and he is bathed in the glare of sensor-activated security lights. To get in, he must unlock a deadbolt, deactivate an exterior alarm in half a minute, then flip a switch to disable further alarms in the floors and doorways of the house. His parents' home, moreover, is one of the least secured in a neighborhood dotted with spike-topped fences and patrolled by private guards. Across Los Angeles, Flusty distinguishes five species of "interdictory spaces" that exemplify our sense of physical insecurity: "stealthy" or obscured; "slippery" or largely inaccessible; "crusty" or walled/gated; "prickly" or uncomfortable; "jittery" or actively monitored. You've seen these places: the pocket park behind a fence, the sharp-edged garden fence, stores ringed with cameras. Such spaces work to "intercept and repel and filter would-be users"—whether they belong there or not.

> That first night, she started to get out of bed. I believed we should go in right away. My wife was of the belief that we should let her explore, see what it's like to be alone in her room in the dark. We went in. We're conditioning her to stay in bed. We watch when we put her down for a nap, until she goes to sleep. I had it wired up over the weekend. I was watching the game but I had a small box in the left-hand corner of my TV. My wife, if she wakes up in the middle of the night, will flip it on. After a week, Zoe did ask what it was. My wife just told her in very simple words that it's a camera and that it allows us to see into her room.

The invention of the Panopticon marked the "point of explosion," says philosopher Michel Foucault, when societies built upon webs of social management became the norm.[12] Whereas kings once ruled by war and torture, by the late Middle Ages subtle and indirect means of exercising power began to spread, capillarylike, into every nook and cranny of society. Thickets of regulations shaped behavior. Nearly every act became supervised and recorded. Prisons, police, hospitals, asylums, and public schools developed as nodes in modern "disciplinary" societies, according to Foucault. And surveillance was the essence and ideal of such cultures. The major achievement of Panoptic control, writes Foucault, is "to arrange things [so] that the surveillance is permanent in its effects, even if it is discontinuous in action; that the perfection of power should tend to render its actual exercise unnecessary; that this architectural apparatus should be a machine for creating and sustaining a power relation independent of the person who exercises it; in short, that the inmates should be caught up in a power situation of which they are themselves the bearers." As Foucault notes, the real danger of the system lies not in its repressive qualities but in its ability to mold people.

Yet what was once solid is permeable and what was once grounded is mobile. Institutions and traditions of the past are fraying. Now we live in societies of *control*, not discipline, marked by systems with ever-shifting perimeters like "a self-deforming cast that will continuously change from one moment to the other," argues philosopher Gilles Deleuze. "The man of control is undulatory, in orbit, in a continuous network. Control is short-term and of rapid rates of turnover, but also continuous and without limit."[13] In other words, surveillance is free-floating, operating not just in closed institutions such as homes, offices, and schools but in all spaces. Venture into a new corner of the Web and ping!—an e-mail triggers a day of parent-mandated cyberspace exile. An "event data recorder" tracks a teen's driving speeds and braking style, while a bumper sticker asking "How's My Teen Driving?" exhorts strangers to join in the surveillance. GPS on a cell phone transmits a child's every location. Browser histories, data printouts, and camera footage capture it all. In effect, we've moved beyond the idea of the potential gaze to something closer to an ideal of unblinking vigilance. "I don't use it like a spy," Peter Kleiner says of the surveillance system in his New Jersey home. "I use it for knowledge."[14] Kleiner has sprinkled seven cameras around the house, from the garage to his bathroom, to track repairmen and cleaners secretly and to check up on his thirteen-year-old and fifteen-year-old sons, with their knowledge, when they come from school or play with friends in the basement game room. He can referee a sibling spat because he saw it all on camera, or sit in his kitchen, tele-chaperoning a gyrating teen party downstairs. "I'm just one of those parents who wants to know what goes on," says Kleiner, the manager of a Philadelphia radio station. "Knowledge is power."

Blanket surveillance, however, doesn't just catch transgressions. The gaze reveals and records the mistakes, the possibilities, the foot stretched gingerly over the threshold of what's allowed. The goal is an inescapable level of deterrence—to catch the potential not just the actual transgression. "Imagine if you could sit next to your teenager or employee every second. Imagine the control you would have," writes Larry Selditz, president of a maker of vehicle surveillance systems.[15] School districts now offer parents online class attendance records, assignment-by-assignment grades, report cards, and readouts on every cookie or double cheeseburger on the lunch tray. This is what David Lyon calls "actuarial justice," in which the prediction of deviance is as important as the catching of the crime.[16] We are not living in the soma-induced docility of *Brave New World* or among Orwellian terrors of *1984*. And yet we have crossed a certain threshold when we—often unthinkingly—create such webs of surveillance *for our own kin*. Who's taking a cookie, pushing the speed limit, wandering the Web, climbing out of bed? Who's tiptoeing over the edge of the safe zone?

> At first she knew she wasn't supposed to get off that bed. Now she knows she can, and there aren't going to be dire consequences. If she stands on the footboard and reaches for the light switch, that's when

we'll go in. . . . . I consciously think about how I'm going to intervene in certain situations. I'll let her go all the way to the point where she's going to hurt herself or get paint or ink on something before I intervene. I want her to make good decisions. We're trying to equip her with the right decision-making abilities. She has a sign in her room, "Well-behaved women rarely make history."

In one still shot from ZTV, Zoe is curled up on her side, her footed pajamas glowing eerily in the dim, gray room. Her back is to an unadorned wall and her tiny body is partly wrapped in a blanket and tucked into a corner of her seemingly enormous bed. She is asleep, but the camera is watching over her.

I captured a little video and a couple of screen shots just for our "Zoe Journal," so that someday when she's 18, I can joke with her that I've got cameras hidden all over the house.

The crowd milling about the exhibit hall didn't take much notice of the dozens of small vials neatly lined up at a research laboratory's booth. Semtex, ammonium nitrate, smokeless power, dynamite—perhaps these "traditional explosives" are passé. Two police officials from the United Arab Emirates' Interior Ministry paused, befuddled but fascinated by tiny pocket screwdrivers being given away at the booth. After a moment's hesitation, they hurriedly took several, as a perky sales rep warned, "You can't take them on the plane!" Most attendees at the Fourth International Aviation Security Technology Symposium, a US government–hosted event held every five years, were drawn to the hulking, chalk-white machines scattered around the ballroom of a Washington, DC, hotel. These machines are the top guns of borderland security: the screeners for baggage, irises, cargo, and fingerprints. They are the state-of-the-art vision machines that augment but can't yet replace the miracle that is the human eye.

We are a visual species. We evolved to take in the world through our eyes, probably because understanding the environment was the best hope for our survival. Our eyes are the "great monopolists" of our senses, writes poet Diane Ackerman, the information gatherers par excellence. "To taste or touch your enemy or your food, you have to be unnervingly close to it. To smell or hear it, you can risk being farther off," she writes. "But vision can rush through the fields and up the mountains, travel across time, country and parsecs of outer space and collect bushel baskets of information as it goes. . . . For us, the world becomes most densely informative, most luscious, when we take it in through our eyes."[17] Moreover, in the name of reason and science, we have grown even more enamored with and dependent on what Yi-Fu Tuan rightly calls our most intellectual sense, even to the point of letting our other senses lie virtually fallow. "From the sixteenth century onwards," he writes, "writers paid greater attention to what is and is not visible from a certain standpoint and to consistency in the relative size and scale of objects in a scene."[18] In the eighteenth century, a doctor could still diagnose diabetes, kidney failure, and other illnesses by smell, but over time, such sensory prowess faded as the eye became the arbiter of truth. "Observation rather than the *a priori* knowledge of medieval cosmology was viewed as the basis for scientific legitimacy, and this subsequently developed into the foundation of the scientific method of the West," said sociologist John Urry.[19] Today we firmly depend on our powers of visual attention to understand our environment. The eyes have it, we've come to believe.

But do we really see all? How perfect is the gaze? The aviation security conference, where some of the country's best visual attention scientists gathered, gave an alluring peek into just how difficult it is to see the world, even when you are trying hard to do so. Think it is easy to spot a gun or a knife in a suitcase passing through airport security? Consider that after twenty-four hundred chances, people are not much better at it than when they started, University of Illinois scientist Jason McCarley told a packed "human factors" workshop at the four-day conference.[20] After practice, his test subjects halve the number of times—from 20 percent to 10 percent—that they look right at the weapon but fail to recognize it. But they still fail to locate the weapon *at all* on nearly a third of their tries. Tell them to speed up, and they make more mistakes. Coach them to take care, and they slow to a crawl although continue to make frequent errors. Why? The problem, of course, is not strictly visual. Recall that attention is a kind of spotlight, highlighting a "coherence field" of perception from among neuron groups competing to represent sensory information to the brain. Attention casts the deciding vote in what we perceive of the world and so is the beast to harness in a world of cognitive overload.

A big part of the problem for baggage screeners is the rarity of the "prey," scientists are discovering. Finding a gray gun in one bag doesn't help you spot a brown-handled knife—different shape, size, color—in the next, because the experience of finding one seems to differ greatly from the next. "There's little opportunity for top-down control and learning, a poor transfer of skills," said McCarley. We also tend to be hardwired to quit looking for rarely seen objects—an instinct that evolved to promote optimal foraging and other survival skills. Put a "weapon" in 50 percent of suitcases in laboratory experiments using computer simulations, and volunteer observers miss them just 7 percent of the time, reported Harvard Medical School professor Jeremy Wolfe, a wiry man with a dry sense of humor. "Booby-trap" 1 percent of bags and nearly a *third* of the weapons will be missed. Like most such research, Wolfe's study used volunteer test subjects, not professional scanners. Still, the results are "grim," Wolfe told a sea of stony-faced security experts.[21] Wolfe's suggestion? Take screeners off the line often for booster shots of training. A nervous researcher from England concluded her talk with another interesting suggestion: have one screener look for guns and another for knives, since we can't seem to search for more than one target at a time. By day's end, I was silently thankful to be taking the train home.

Our daily perception of the real world is likely no better, as evidenced by a famous one-minute movie that sets up one of the most striking psychology experiments of recent times.[22] The film, which scientists of attention love springing on the uninitiated, shows six people, half dressed in light clothing and half in dark, moving around a room while passing two basketballs back and forth. Told to count the number of passes made by one team, half of viewers completely fail to notice a woman in a gorilla suit who calmly walks through the group, briefly pausing to pound her chest. All told, she is on screen for nine seconds—an eternity in attention research. "Nothing undermines the confidence in what you see with your own eyes like the opaque gorilla video," marvels science writer James Gorman, reporting on one of the field's many iterations of the study, in which just 18 percent of mildly inebriated people noticed the ape, compared with 46 percent of sober viewers.[23] "We were doing this on a lark," admits Dan Simons, a University of Illinois scientist who set up the original experiment in 1999 as a project for his undergraduates. "We didn't think it would work."[24] The concept of "inattentional blindness," that is, missing something fully visible that we don't attend to, had long been recognized yet never fully appreciated until first Arien Mack and Irvin Rock in their landmark book *Inattentional Blindness* and later Simons realized its significance as the emperor's new clothes of cognition. Now a whole genre of studies has sprung up around studying this and other failures of awareness. In one of McCarley's experiments, adults talking via microphone with someone in another room while simulating driving frequently failed to see crucial changes in traffic scenes, such as a child darting between cars, although they were looking directly at the pertinent area of the scene.[25] Other research reveals that we are effectively blind during the thirty to fifty milliseconds it takes to move our eyes, so counterintuitively, the more you look, the less you see.[26] Far more than a cocktail party trick, the "Gorillas in Our Midst" movie is a window on how much we may be missing in a world of rising overload.

But what's most unnerving and fascinating to Simons is how cocksure we are of our ability to take in the world. Told that they'd missed the gorilla, quite a few test subjects refused to believe of its existence until shown the movie a second time. Why? "We think of our eyes as video cameras and our brains as blank tapes to be filled with sensory inputs," observes *Scientific American* columnist Michael Shermer in an article on the gorilla movie.[27] Simons adds that the "richness of our visual *experience*" fools us into thinking we've captured it all—for automatic playback at anytime. Most of us are far more aware of the world than simultagnosiacs, who suffer from an extreme form of attentional neglect and can only perceive one object at a time. Still, we aren't the all-seeing species that we think. Perception of all kinds is really a construct and so only a rough and incomplete snapshot of the world. "The body edits and prunes experience before sending it to the brain for contemplation or action," says Ackerman. "This makes our version of the world somewhat simplistic, given how complex the world is. The body's quest isn't for truth, it's for survival."[28]

And yet the urge to see all, to probe, and to capture some form of "truth" drives us onward in the era of the panoptic gaze. With ever-more potent machinery, we piece together the *construct* of one another that we hope will ultimately lead to a rich, full knowing. Surveillance is both a tether and a portrait, made bit by bit from pieces of me and of slices of you. Click. We've bested Bentham's fallible warden, turning round and round in his lonely tower with tired eyes. We have enlisted the camera, and all its wondrous iterations. Click. We *are* the camera.

> In a way, nobody sees a flower really, it is so small,
> we haven't time—and to see takes time, like to have
> a friend takes time.
> —Georgia O'Keeffe[29]

Bathroom behavior. Shouts and gestures of motorists. Female taboos about eating. The private lives of midwives. Which end of a cigarette people tap before lighting up. The marching orders of Britain's prewar Mass Observation movement were eclectic, at best. Created in 1937 by a painter, a poet, and an anthropologist to study minute details of working-class British life, the effort flourished for just eight years before splintering, but it serves as a sometimes-comical illustration of our often-blind faith in the power of observation. "Only mass obser-vations can create mass science," poet Charles Madge wrote in a call for recruits to the nascent movement.[30] For the May 12, 1937, coronation of George VI, Mass Observation collected forty-three volunteers' diaries, seventy-seven questionnaires, and the observations from twelve observers. In this way, varied "kinds of focus were obtained," explained Madge and cofounder Humphrey Jennings, a surrealist painter and documentary filmmaker. "Close-up and long shot, detail and ensemble."[31] The movement's army of observers, who wrote about their own lives as well as those of strangers, were not just witnesses to history, collectively detailing "weather maps of public feeling." They were supposed to *be* cameras, offering "continual observation." In 1961, Madge wrote that he felt thankful that "some sort of a net had been spread to catch that fleeting, glinting apparition: the essence of the time."[32] True, the gossipy, random snippets of life depicted in the movement's carefully preserved archives do create a mesmerizing and strangely timeless collage of human experience—an "essence" of a sort, no doubt. (From housewife Phyllis Walden's diary of September 12, 1937: "8:30. He asks, 'what's for breakfast?' I say, 'bacon, fried potatoes and fried bread'. Says he'll have his in bed. Very glad about this, as table very small when we all meal together and he does find fault with the children's table manners so.")[33] What's more riveting is the fact that Mass Observation's undisciplined and largely unscientific creators, their observers, and even their subjects so cheerily placed themselves before the watchful eye of the movement, believing that they could somehow capture a complete and unblinking portrait of their days. "Mass Observation sets out to be a new science: or rather a new method of finding out scientific truths," gushed magazine writer Marion Dewhirst in June 1938.[34]

What were they striving for? Perhaps they sought the impossible: a full portrait of humankind. And just as we confidently assume that our naked eye takes in the entire landscape before us, so we believe that a photograph will not lie. A photograph is borne of the augmented eye. What more powerful portrait of life can there be? "If we extend our eyes by attaching artificial lenses and other accessories to our real ones (glasses, telescopes, cameras, binoculars, scanning electron scopes,

CAT scans, X-rays, magnetic resonance imaging, ultrasound, radioisotope tracers, lasers, DNA sequencers, and so on), we trust the result a little more," muses Ackerman.[35] In fact, we trust the result much more, but in doing so we mistake new views of the world for new whole truths. A photograph, like an eyewitness report, provides visible proof and so is accepted as evidence, writes Susan Sontag. As the product of a machine, a photo also carries an aura of objectivity and "innocence," she writes. But we tend to ignore that photographs, like paintings, are interpretations of reality. "Humankind lingers unregenerately in Plato's cave, still reveling, its age-old habit, in mere images of the truth," writes Sontag in her essay collection *On Photography*.[36] She offers a piercing depiction of how the camera—the foremost of our surveillance technologies—has changed the way we see the world.

In the age of the camera, reality becomes "a series of unrelated, freestanding particles," an inventory of fragments. "Photography implies that we know about the world if we accept it as the camera records it," says Sontag. "But this is the opposite of understanding, which starts from *not* accepting the world as it looks." Real understanding unfolds over time, as Georgia O'Keeffe observes, whereas the camera offers "knowledge at bargain basement prices—a semblance of knowledge; a semblance of wisdom; as the act of taking pictures is a semblance of appropriation."[37] Photographers, who operate within the surrealist sensibility that inspired Mass Observation's founders, "suggest the vanity of even trying to understand the world and instead propose that we collect it," asserts Sontag. As a result, life becomes a trail of information, not a sense-making river of experience —a criticism fired at Mass Observation by some contemporaries. "The facts simply multiply like maggots in a cheese," one reviewer complained of the movement's slim book about the coronation, "May the Twelfth: Mass-Observation Day-Surveys 1937 by over Two Hundred Observers."[38] The end result: continuities are fragmented and fed into an "interminable dossier, thereby providing possibilities of control that could not even be dreamed of under the earlier system of recorded information: writing," according to Sontag.

. . . when she's 18, I can joke with her that I've got cameras hidden all over the house.

Alice Byrne has seen the dossiers. She is a veteran New York private detective, and hiring her is a last resort for an anxious parent. Three decades ago when Byrne stumbled into this work after a few years in psychiatric nursing and a few in security, she never heard from parents. Now they come, loaded with hoarded evidence, looking to Byrne to turn the fragments into an understanding of their child. "They'll lay out things on this table, the records they have," said Byrne, sitting in the dining room of her home in Newkirk, a working-class Brooklyn neighborhood of stolid houses with postage-stamp lawns.[39] Days before I visit Byrne, a mother had arrived with a three-inch stack of documents on her seventeen-year-old: cell phone, text message, e-mail, and credit card records, plus her handwritten notes on his comings and goings. "That was her amateur sleuthing, but she didn't have the answers she needed," said Byrne, a sixty-something grandmother with shoulder-length blonde hair and a comforting gentility who nonetheless rattles off tales of bodies in rivers and tough-cop interrogation tactics once she warms to a visitor. "She had no idea."

Byrne no longer goes undercover. Multiple sclerosis keeps her in a wheelchair, directing a squad of ten detectives by cell phone and instinct, a spymaster supervising domestic wars she cannot see. But she knows that in her business, nothing can substitute for street work—and time. Printouts don't tell a story. GPS—it's just a dot on a map. "Okay, they're there, but who are they with, what are they wearing? What are they asking? What are they saying? You can't use technology to do that." Tailing a subject for at least a week and up to a month is needed in order to get a feel for their lives. When she closes a case, Byrne always writes her reports "like a story," she said, pausing to sip from a mug of tea. "It's not going to say, 'four o'clock, subject entered this place.' It's going to talk about how they dressed, how they looked, how they walked, things that might mean something to someone who knows them, because they can say, 'They never walked fast in their life, do you know,' 'I never saw them with their hair down!' or whatever. You try to paint a picture in the report that will help answer questions along the way."

Surveillance is a business of fragments, a language of traces. In a distanced, crowded world, we must depend on innumerable symbols and relics, from driver's licenses to fingerprints, to display our legitimacy and rout possible deviance. Thousands of years ago, Babylonian merchants and Chinese officials used fingerprints as signatures on documents, according to Robert O'Harrow Jr.[40] In 1892, an Argentine police inspector used fingerprints for the first time to solve a murder; in that case, to show that a mother had killed her two sons. Now, the FBI keeps millions of prints on file, and both companies and law enforcement agencies worldwide collect masses of other biometric records based on faces or irises. There are pieces of you sprinkled all over the Web—from your Facebook page to the number of times you've visited the Victoria's Secret site—and digital profiles are offered for sale in a kind of invisible alter-universe/marketplace. And then there are the cameras: seven million by one estimate in the United Kingdom, and hundreds of thousands proliferating across the United States. Chicago mayor Richard M. Daley calls the cameras he's ordered in public spaces "hundreds of sets of eyes."[41] This massive web of data seems ordinary, even comforting, allowing us to gain entrance to slippery, prickly spaces, to maneuver complex routes and get the goods we need to survive. It is our ether—until something goes wrong. Johnnie Lockett Thomas, a seventy-one-year-old African American from Montana, is detained every time she flies because her name is similar to John Thomas Christopher, the *alias* of a white man in prison for murder, O'Harrow reports.[42] The info-bits of one have melded with the other, subsuming the elderly widow, at least for the frightening moments she stands before airline security. As well as becoming invisible, the object of the panoptic gaze is silenced. "Newer regimes of . . . surveillance and security require less and less that the subject be—literally—response-able," notes Lyon.[43] Our databit-and-snapshot portraits are drawn and redrawn in codes that we do not comprehend. With one tainted trace, we can be quietly erased.

And when we depend on a collage of fragments, an "interminable dossier," to stand for the whole person in private life, the cost is higher. We cannot gain an *understanding* of those closest to us through traces—computer printouts, GPS locations, shots from a hidden camera. The spouses in the Technology Plays misread one another completely based on scraps of evidence. The one who is duped still has faith, while another who's been virtually cheating on his wife thinks erroneously that all can return to normal. We think we see all, but we fall poignantly short, even with our augmented eyes. (What is Alice Byrne to her clients but an augmentation, like spyware or a data recorder, hired to enhance their gaze?) The incompleteness of this knowledge is one reason why surveillance efforts never seem sufficient, and parental efforts to monitor their children so often contain a whiff of desperation. One more peek at the baby-cam in the middle of the night, one more glance at the teen's blog, then, we'll know for sure that they are safe—or guilty. "As soon as you have one camera, you want to see around the next corner," Lyon quotes a colleague as saying. "Then you want to see in the dark as well, so you add infrared cameras. Then you want to see more detail so you add more powerful zoom lenses. And then you think, if only I could hear what they are saying as well, so you add powerful directional microphones to your cameras."[44] Ultimately, the insatiability of the omnipotent gaze led Mass Observation to slip from observation into manipulation. Anthropologist Tom Harrisson began secretly reporting on wartime morale for the government, angering poet Charles Madge, who objected to the movement's slide into "home-front espionage." Painter Humphrey Jennings directed wartime propaganda films. In 1949, Mass Observation became a market research firm.[45]

Alice Byrne finished her tea and took a call from a detective on the street. She is a weaver of snippets, and she knows how incomplete even her carefully constructed narratives are and how toxic a bit of information can be. Years ago, she put a psychologist on staff, because she felt that clients—even those whose suspicions were unfounded—were unsure how to move on after a case closed. "You give people things that change their life, and it always seemed to me, 'okay, you did your job,' but there's more to it than that," she said. There's a river of experience beneath the trail of fragments that clients vigilantly collect. As our conversation wound down, she mused about raising children in a surveillance society. "If I thought I was being watched all the time and I had no privacy whatsoever, I would really begin to resent whoever had done that to me first of all," she said, driving her wheelchair to the door to see me out. "And I would become more secretive." She shook my hand and turned to take another phone call. As I left, I glanced back at the pumpkin and string of Christmas lights decorating her front porch. A warm, yellow light spilled from her windows into the gray dusk of a winter afternoon.

In June 1983, French conceptual artist Sophie Calle found a man's address book on a Paris street and photocopied it. She interviewed those listed in the book at length about the owner, then published a twenty-eight-day series of articles about him in the national newspaper *Liberation* in August and September, accompanied by photographs of things connected to his life. "Through [his friends], I would get to know this man," Calle later wrote. "I would try to find out who he was without ever meeting him and produce a portrait of him, . . . which would depend on the willingness of his friends to talk and the turn taken by events."[46] Apparently, they were more than willing, for her mesmerizing portrait of Pierre D., however embellished by the artist, is excruciatingly precise. He is a man who "keeps a file on his enemies in his bookshelf entitled 'My Hates,' . . . wears sagging clothes, . . . is always ready to fall in love as long as he hasn't a chance, . . . is organized in his solitude . . . Mysterious. Someone who would be capable of disappearing without a trace."

The "Address Book" project was in keeping with Calle's artworks, which she calls "investigations." She has worked as a maid in a Venice hotel so she could rifle through the possessions of its occupants, photographed sleeping people, and followed strangers "for the pleasure of following them, not because they particularly interested me," she recalls. "I photographed them without their knowledge, took note of their movements, then finally lost sight of them and forgot them."[47] Her appropriation of intimate, unwatched moments, often through photography and the ritualization of life by, for instance, exhibiting years of her birthday gifts, stems from a deep yearning for "a well-ordered life," writes *New York Times* reporter Alan Riding.[48] Certainly, Calle's acts of surveillance give her a subversive control over others. Twice over a span of two decades, she has had herself tailed by an unwitting detective, an act that turned the "all-seeing" private eye into a fool led on a Paris goose chase. But Calle's art does more than exert control, as philosopher Jean Baudrillard has written. "Please follow me," he appeals to Calle in an essay on her work. Baudrillard argues that through the act of shadowing, both the follower and the one pursued effectively become shadows and disappear.[49]

What does this mean? The answer lies in what Calle's work ultimately shows us about surveillance and trust. Calle controls the parameters and outcome of her every encounter—from a panoptic fling with a stranger to the appropriation and exhibition of intimacies with friends and lovers. She's the queen of risk management; the other always winds up in the chokehold of her gaze. In contrast, trust is a *risk-taking* that allows for the freedom and agency of another. As Russell Hardin explains, "trust is embedded in the capacity or even need for choice on the part of the trusted."[50] If there isn't at least the possibility for choice, even for a potential betrayal, there can be no trust. To trust is to be vulnerable, says philosopher Trudy Govier.[51] In such a relationship, there is always something at stake—the future of the connection, a privilege or consideration. By trusting someone, we also acknowledge his or her capability to be trusted in some respect. Baudrillard was right. In a sense, we do erase those we watch, because paradoxically while making them visible, we are imprisoning them. Finally, perhaps we, too, fade into the darkness of surveillance, invisible wardens in our lonely towers.

Surveillance suffocates trust in another way. Trust isn't instantaneous but happens gradually as the result of small-scale interactions, continual assessments, and a "rich understanding" of the other, says Russell Hardin, a slight, pale New York University professor with a patient air who is an international expert on trust. "Familiarity gives us *knowledge* about the other, in particular about the other's trustworthiness," he says. All this contrasts, of course, with the "semblance of wisdom" we gain from amassing a thick file of info-traces. "Distrust comes easily because it can be built on a limited bit of behavior," says Hardin.[52] When I e-mailed him to ask for an interview, he instantly proposed lunch. How trusting! We met at his office and then walked to a nearby Italian restaurant, where he reminisced about the mozzarella he had tasted recently in Florence, offered a few observations on trust, asked about my work, and gave me one of his books with a parting peck on the cheek. We even traded stories about relatives who no longer speak to one another. Still, I left somehow dissatisfied, thinking I was going to get something concrete, a good quote at least. Later, I realized that he was not withholding but rather treating this unhurried lunch as a beginning, suffused with possibility for further exchanges, while I had approached it as a one-off interaction. Trust is an openness to something more, distrust a shutting down. Hardin is a trusting soul.[53]

Recently, I met a friendly French man at a neighborhood bakery. He was moving to New York shortly with his wife and two children and we traded e-mails, then got our families together—minus my teenager—one afternoon in the park. To my surprise, they asked if my fourteen-year-old could babysit the next week for an evening. After one brief encounter with us, they were ready to take a risk on my daughter. We took a much smaller risk entrusting her to them, amid much anxiety on my part. It all worked out, but I couldn't help wondering whether our society has raised the bar for trusting others so high that we perpetually miss out on some essentials of life, from serendipitous encounters to social capital or even just the ability to do good. On Christmas Day, my sister-in-law's wallet was stolen at the Providence airport at the beginning of a much-delayed flight home to North Carolina. Discovering the loss at 2:30 a.m. during a layover, she asked a seatmate with whom she'd been chatting for a $15 loan to get her car out of the airport parking lot at home. He refused. The same week, our new French friend tipped my husband to an important business opportunity.

If you generally doubt people's trustworthiness, you aren't likely to take risks on cooperating with or even knowing others further. This is why trust, once lost, is so hard to rebuild, while distrust breeds quickly and easily. "Low-trust" couples in studies compiled by Hardin "pull back" from dealing with contentious issues, so they miss out on opportunities to show concern and caring. The less care they show, the more effort they make to shield themselves against the breakdown of their increasingly fragile relationship.[54] This is a turtle's view of life, as illustrated by cultures of pervasive distrust around the world, where cooperation is minimal or absent. In the Middle East, traditional Omani households live in cloistered compounds, and interactions between dagger-wielding men outside these walls are highly scripted. Edward Banfield's classic study of a mistrustful southern Italian farming village in the 1950s reveals a society nearly devoid of social cooperation. Even within families, tensions ran high and life was glum. Banfield recorded 320 anecdotes of daily life and folk stories, notes Trudy Govier, but only a few were happy.[55]

Raising our children under the telescope of our permanent gaze is costly. If you are trusting, you win some encounters and lose others, but in the long-term, you gain much more than you would from distrusting, which results in lost opportunities, says Hardin. A panoptic culture teaches our children that we cannot take a chance on others. Unintentionally, such a culture also teaches children that they are to be distrusted, and that we cannot take a chance on them. Cameras, breathalyzers, software monitoring, GPS tracking, and other far-reaching paraphernalia of the eye take away our children's brief chance in life to gradually, with inevitable stumbling, learn to take responsibility for their actions. Surveillance erodes their freedom to fall. "What I always say to parents is if you are giving your kids appropriate freedom, it will feel like neglect in our culture," says psychologist Wendy Mogel, author of *The Blessings of a Skinned Knee*.[56] We're setting up safe zones that are cages. And we're substituting instamatic fragments for the homegrown mutual knowledge that slowly builds into the soulful gamble called trust. We've mistaken the monologue of surveillance for the dialogue that is care.

To recover from this loss, we have to learn when to look away and when to meet each other's gaze. In many ways, knowing when and when not to watch one another, physically and socially, is a difficult dance of attention. Trust, concludes Hardin, is really a "capacity to read the commitments of others, a capacity that must largely be learned."[57] Consider the face-to-face encounter. We are naked and vulnerable before the totality of each other's senses, sharing a "special mutuality" of interaction, according to sociologist Erving Goffman. "Each giver is himself a receiver and each receiver is a giver."[58] Fallible as it is, the eye is the wondrous centerpiece of this contact. Babies begin to learn about the world by following and then *sharing* the gaze of another, as a first toehold in the complex waters of relating. Throughout life, "civil inattention" is sometimes warranted, while at other times looking away is a breach of the social flow. Eye contact is a sign of respect, yet held too long, it can become a "hate stare" such as those that Southern whites gave to blacks in the 1950s, Goffman observed.[59] Through physical and verbal cues, we learn to indicate our "involvement allocation"—whether we are "in" or "out" of play in the encounter. Essentially, our behavior in a moment of presence is an "idiom for expressing attachment," that is, for showing our capacity for paying attention to one another.[60] It is the flip side of surveillance. In tracking another, we do not *exchange* the gaze. Surveillance returns us to the question of the "disappearing bodies," and the dwindling value we place on the earth and our humanity. Surveillance is not holding us together but rather ensuring that we remain apart.

Trust created from presence is the answer. "Trust—in a person or in a system, such as a banking system—can be a means of coping with risk, while acceptance of risk can be a means of generating trust," writes Anthony Giddens.[61] If we are to create a joyful society rich in cooperation, we must especially liberate our children from the panoptic gaze. Children need space to experiment in a world of mutable selves, relations, and institutions, and they need to be given the chance to build what Giddens calls "active trust" born of mutual disclosure. Without safety nets of trust or secure traditions, he says, relations crumble into obsession and compulsion. Or revenge. When Pierre D. recognized himself in Calle's exposé, he sent the newspaper a picture of Calle naked, which was published with her head cut off. Calle later conceded that her investigation had gone too far.[62] In a society built on trust, we may never fully know one another, but we will be given the opportunity to reveal ourselves willingly, over time. We will be given the chance to open up, the opportunity to create something together based on love and respect, not force.

Jeremy Bentham was a visionary, but his scholarly genius came at a steep cost. Tenacious, self-reliant, and egotistical, he led a reclusive life, "reducing the law to a system and the mind of man to a machine," wrote William Hazlitt, an eminent scholar who knew Bentham.[63] In both his private life and his work, Bentham showed an inability to understand the emotional depths of human nature and an obsessive devotion to his own interests and routines. The philosopher demanded near-worship from friends, cut off relationships at a whim, and delivered long monologues to visitors, whom he would see only one at a time. "His view of the human mind resembles a map, rather than a picture," wrote Hazlitt in an 1825 biographical essay on Bentham. "The outline, the disposition is correct, but it wants coloring and relief." Bentham fixated on the parts, but not the whole of life, and especially failed to understand that others might have a different view. Lacking "theory of mind," he believed that both men and society needed to be shaped into precise machines; his Panopticon exemplifies these ideals. All forms of joy or spontaneity had to be stamped out in Bentham's ideal world, writes C. F. Bahmueller. "Both were too dangerous to be sanctioned in a world so devoted to the elimination of contingency."[64] Considering this constellation of traits, forensic psychiatrist Philip

Lucas and forensic psychologist Anne Sheeran argue that Bentham may have had Asberger's syndrome, a form of autism associated with high cognitive functioning.[65] For autistics, who have trouble making eye contact, empathizing, and reading others socially, no punishment could be more dire than panoptic surveillance. In both his life and work, Bentham's gaze could not be returned.

Chapter 6
JUDGMENT
Book and Word on the "Edge of Chaos"

John Bidwell unlocked a file cabinet and removed a small box covered in burgundy linen. Bidwell, curator of the printed book at the Morgan Library in New York, spread a velvet cloth on his desk, opened the box, and removed a curious palm-sized book covered in elaborate gray-green floral embroidery. The gilt-edged book is made up of two volumes bound *dos-à-dos* or "back-to-back" style. One side is a New Testament, published in Edinburgh in 1633 in a condensed but lively typeface. Flipped over, the second book opens to reveal the Psalms, published in London in 1635. Originally carried in purse or pocket most likely by a merchant's wife or noblewoman, this twinned book of piety would have been read during her daily devotions. The tiny text itself was handset by a compositor whose knowledge and dexterity were crucial to the flourishing early book trade. A compositor, wrote printer Joseph Moxon in 1683, should be fluent in Latin and Greek and something of a "scholler."[1] In composing, he would place metal letters one at a time onto a composing stick held in his left hand until it held eight or nine lines, then transfer the text to a galley. At the same time, he would "discern and mend"—edit the prose as he saw fit. The result was a cross between a handmade and a factory-built object. No two volumes in London's bustling one hundred and fifty bookshops of the time were exactly alike, to many an author's chagrin.[2] The book, long our most important vessel of language and hence of our attention, was then a work of art, still emerging from its genesis as a manuscript. And now it is a museum piece, brought out from a dark subterranean vault here in one of the world's greatest rare book libraries to repose quietly next to a sleek desktop computer that can thrust its noisy, wriggling, multicolor innards at us with the touch of a key.

Bidwell invited me to hold the centuries-old Bible, and I did so warily, fearing it would crumble dustily in my hand.[3] He assured me with some amusement that even a fifteenth-century book will last longer than many made between 1860 and 1960, simply because of the quality of the paper. Bidwell is a tall, thin man who records his acquisitions both by computer and by hand in a century-old collection of ledgers begun by his predecessors at financier Pierpont Morgan's private library, now also a museum. A faintly Dickensian character, Bidwell chooses and pronounces his words carefully, often drawing out a bon mot with great gusto, followed by a pause or quiet sigh, as when he's ruing the "bloodless" text of the Victorians. Yet Bidwell is anything but a pontificator. He loves to dig into the depths of the English language or the murky provenance of a *codex* artwork and come up with a date or a story that is a link in the chain of print history. Once charged with sleuthing out how a trove of leftover vintage Oxford press paper had been used, Bidwell tracked which blank sheets had been fodder for William Morris's Kelmscott Press and which went into the making of generations of art or textbooks, just from studying the color, weight, "rattle," and watermarks of the old paper. "It was one of the greatest challenges in my scholarly career," said Bidwell with quiet satisfaction as we made our way downstairs to tour his current exhibit on Victorian best sellers.

The museum was closed that day, the hallways hushed and churchlike, and I found myself wondering whether this cloistered haven portends the future of the book. Will the book become an emblem of times gone by, exhibited like butter churns and chain mail, to show youngsters what we've moved beyond? Will scholars and bibliophiles alone speak its language and probe its mysteries, while the rest of us contentedly graze on a steady diet of Web-info? The book isn't dying, asserted Bidwell, wondering just a moment later whether college freshmen still read books for pleasure. Surely, the 174,000 books published in a recent year in the United States mark a literate society.[4] Yet only 30 percent of college graduates can understand a simple document such as a food label, down from 40 percent a decade ago, and nearly 57 percent of Americans don't read a single book a year.[5] The print era, with its crown jewel the book, gave us systems for nurturing unprecedented levels of discourse, analysis, and focus. Where in the free-flowing digital milieu will we find the stepping stones to these higher forms of attention and thought, or even the will to create them? As we plunge into a new world of infinitely connectible and accessible information, we risk losing our means and ability to go beneath the surface, to think deeply. This is the second collective loss, after trust, we nurture in a time of distraction. Perhaps no Good Book kept carefully by the loyal Mr. Bidwell can save us from slipping into a dark age that is ironically marked by the most plentiful stocks of information the world has ever seen. And yet, can we Google our way to wisdom?

"Let go!" a rising chorus chants. The book is surely meant for the museum vaults, if not the dustbins, of history. We've held too long to the notion of the book as sacrosanct, the chosen vessel of the Word and now the failed keeper of a language spilled messily across the universe. In the not-distant future, books will only be cherished by those who are "addicted to the look and feel of tree flakes encased in dead cow," writes MIT architecture professor William J. Mitchell.[6] The writing is on the screen, and we must move on, say some cognoscenti, taking laser-sharp aim at the emblem of the print era: the solitary reader plunged thoughtfully into the recesses of the unchanging text. Books are firstly *unsociable*. Print is an "act of perceptual self-denial," a "decision of severe abstraction and subtraction," a "trickery," asserts rhetoritician Richard Lanham, calling for a new knowledge-making system based "on the edge of chaos." He adds, "Is it automatically more virtuous to inhabit a small, closed world of purely private purpose?"[7] In the networked society, we need not be left alone with ourselves anymore. Next, "liberate" the text! That's the second challenge put to us as we rise up from our creaky library chairs. Books imprison information in rigid and unchanging vessels, like wine stoppered in bottles, says John Perry Barlow. Release the nectar! Information wants to be free, trumpets Stewart Brand. Asks classicist James J. O'Donnell: "Is it not strange that we take the spoken word, the most insubstantial of human creations, and try through textuality to freeze it forever, and again, try to give the frozen words of those who are dead and gone, or at least far absent, control over our own experience of the lived here and now?"[8] The word has been on ice, far from the warm pulsing beat of our dynamic lives.

Such exhortations, as many would-be liberators of the text note, echo those of the Italian futurists, who took particular aim at the rigidity of the book a century ago as they sought to usher in a new intellectual and artistic order. The book is "destined to disappear," along with museums, cathedrals, libraries, and other baggage from the past, predicted the movement's founder Filippo Tommaso Marinetti, who also tried to hurry things along by calling for the burning of all libraries.[9] Writing at the height of the turn-of-the-century fascination with simultaneous art, Marinetti espoused *"esplosivo"* texts that attacked typographic, grammatical, and literary conventions, spewing unpunctuated "free and random" words and pulsing images across the page in a prophetic foretelling of digital possibilities. (He called his free-form aeropoetry "simultaneity tuned up.")[10] Writing in 1913, Marinetti looked ahead to an "earth shrunk by speed" and a "new sense of the world." Man, he wrote, "hardly needs to know what his ancestors did, but he has a constant need to know what his contemporaries are doing all over the world."[11] The futurist leader was rightly discredited by his later association with fascism, notes Lanham, and yet some part of futurism's vision presages our present ambitions. Marinetti was looking ahead to the "operating system of the global village," asserts Lanham.[12]

Snipping ties to the past, however, is neither warranted nor possible. The book, while far less fluid than hypertext, has never been as frozen as some like to think. Writing captured words—which were as invisible in an oral culture as the wind—set them down and raised consciousness and thought to new heights of complexity. Suddenly language had a pause button, and what had been fluid and shared was made stoppable. Plato was appalled. Written words, he wrote, "preserve a quite solemn silence."[13] If you question them, he noted, they simply give the same answer again and again. His critique sounds naive, but that's only because we are so steeped in literacy that we can't imagine the shock that writing imposed on oral cultures based on memory, discourse, and the interpretation of natural symbols, such as landmarks. Nevertheless, the relative stability of writing has not precluded change. Books have evolved to reflect our expressive needs and will continue to do so. From the thousands of *volumen* (scrolls) stored at the ancient library of Alexandria, to the first paginated bound manuscripts of AD 300 or 400, from Victorian "chapbook" serials sold door to door, to modern letterpress works and to Marinetti's combustible creations themselves, books have been shape-shifters, revered for their versatility.

And so, too, has their text flown free, often to a reader's or writer's distress. Along with edit-happy compositors, many medieval scribes, translators, and printers mucked around with the works they were copying, transforming, or publishing. "For a long time, the text, although written, was perceived as a penetrable and modifiable entity by other individuals apart from the author," notes Raffaele Simone, a scholar of linguistics.[14] It wasn't until 1760 that the first book reputedly printed without error was published, asserts book historian Adrian Johns.[15] Authors from Martin Luther to natural philosopher Robert Boyle were beaten to publication by unauthorized copies of their own works. Editorial and legal systems were slowly forged to combat "usurpation" and infuse fixity into the publishing process, and yet no age has been able or willing to truly bottle the book and its text. Looking at the Morgan Library's silent, unimposing Scottish bible, we easily lose sight of how many iterations the book has taken—and just how mutable indeed is our relation to it. "We forget that every text is, in a very essential sense, "interactive," changing according to a particular reader at a particular hour and in a particular place," says literary critic Alberto Manguel.[16] In other words, the attention we accord each text essentially creates the book for us, the reader.

Lastly, we often forget that the book is a garrulous beast. In moving into a dominant place in our culture, writing did not categorically turn knowledge-seeking into a still, lonely undertaking. Laced through this vast cultural shift are the sounds of words lifted from the page and spoken in studies, clubs, parlors, school rooms, law courts, and auditoriums—evidence of the long and fertile reach of our oral past. Charles Dickens nearly killed himself at the end of his life trying to satisfy the demand for his readings. Nineteenth-century European cigar makers, tailors, and other craftsmen hired someone to read to them while they worked, or took turns doing so, writes historian Robert Darnton.[17] A century earlier, when French or German peasants gathered in the evenings, children romped, women sewed, and men repaired tools, while one of their number read aloud, adds Darnton. American children studied their wood or even gingerbread "hornbooks," a simple board with a sheet of paper attached, and later, their "McGuffey Readers." Scholarly success meant being a good orator, not a great reader, even a few generations ago. Silent reading arose in monastic scriptoria for religious purposes in the early Middle Ages but did not become routine in aristocratic circles until the fifteenth century. "For most people throughout most of history, books had audiences rather than readers," says Darnton. "They were better heard than seen."[18]

All this goes to show that the versatile, ever-changing book has brought us to the doorstep of the digital age. We are not starting fresh and inventing something uniquely free and fluid by turning to the screen. Relations between book and screen are better described as dynamic, rather than a dichotomy. Both, after all, are communications technologies, which history tells us have a way of messily coexisting, rather than neatly canceling each other out. Painting and cinema, the telegraph and telephone, television and the Internet, trains and planes: all have coexisted despite dire predictions and exhortations to the contrary. "The actual relations between emerging technologies and their ancestor systems proved to be more complex, often more congenial, and always less disruptive than was dreamt of in the apocalyptic philosophies that heralded their appearance," write David Thorburn and Henry Jenkins in *Rethinking Media Change*.[19] There are always "winners" and "losers," such as print and illuminated manuscripts, and more commonly "front-seat" and "back-seat" technologies, such as television and radio. The book may well eventually end up relegated to the museum vault or fade quietly *into* the screen, and yet in our digital infancy, we will undoubtedly experience a messy coexistence of the two for a long time to come.

The real question going forward is *how* we will read, for in this explosively information-laden world, reading—whether a screen, a book, or scroll—lies at the crux of our ability to perceive and make sense of the world. The form or vessel of our text is crucial, but the medium is just *one* of many messages that can be heard in the cacophonous culture of reading. This miraculous cultural invention, a feat so unnatural that it necessitates a painstaking rewiring of the brain to learn, shapes our understanding of life. Done well, reading gives us a deeply meaningful framework for living. Apart from love, it is our only vehicle for attaining "secular transcendence," argues critic Harold Bloom.[20] Done poorly, no matter what we read, we are marooned on the surface of sense-making, eternally adrift. "In these new technological spaces, with these artifacts that will certainly coexist with (and in some cases supplant) the book—how will we succeed in still being able to invent, to remember, to learn, to record, to reject, to wonder, to exult, to subvert, to rejoice?" asks critic Alberto Manguel. "By what means will we continue to be creative readers instead of passive viewers?"[21] The answer lies not just with the text but with the *context* of reading, from the places and times to our means and motives for doing so. We need to consider what kind of literacy we will forge for a society awash in words and yet increasingly lacking the higher forms of attention that form the bedrock of reading.

Back in his office, John Bidwell opened a slim volume to a page of Latin liturgical text reproduced from one of the library's "incunabula," or books from the pre-1501 cradle period of printing. The text is a dense mass of close-knit Gothic script, little interrupted by punctuation or spacing. "You can tell from the typography, from the way the text was punctuated, from the way the words were abbreviated that these texts were meant to be pored over word by word, and read and reread," said Bidwell. "This is not the kind of text that you're going to skim. The act of reading required great concentration." The *Sarum Missal*, a book used in services at Salisbury Cathedral, harkens back to the era stretching from the first Greek alphabetic writings around the eighth century BC, to about AD 1000, when texts were *scriptio continua*—thin columns of uninterrupted letters, initially in capitals. THESEWRITINGSWEREONEROUSTOREAD, as you can see from this brief example, but *reading* as we know it was not really the point. "The written word was more often used in the service of the spoken," writes Rosalind Thomas, a literacy and classics scholar at Oxford.[22] Politics and learning were oral arts, and even literature was invariably spoken or sung. Texts were mostly an aide-memoire, a kind of cue for recalling knowledge that was already well learned orally, much as we consult notes for a talk. Early medieval monks carried out the *ruminatio* of a text, a kind of incorporation through meditation and prayer of the text's wisdom into their faith. For hundreds of years, books were rare and they were deeply known, as familiar as one's hand and as assimilated into one's being and memory as a love is woven into your heart.

This "intensive" era of reading began to be eroded in the mid-1700s by the "reading revolution"—the spread of schooling, book production, newspapers, reading societies, and lending libraries.[23] An early eighteenth-century New England family might have read the Bible, an almanac or two, the *New England Primer*, and Philip Dodd-ridge's *Rise and Progress of Religion in the Soul* for generations. But a century later, they would have had access to novels, newspapers, children's books, and pamphlets, notes Darnton. Intensive reading still took place, but more and more, "extensive" readers, who read "ravenously, discarding one thing as soon as they could find another," became the norm.[24] This was the era of the first reference guides—from train tables and travel books to encyclopedias. Reading became quick and critical, so much so that by 1898, the *Chicago Evening Post* cautioned against the "library habit of reading," in other words, reading without properly "assimilating and digesting" material. Reading quickly, and even "skipping" around in the choice of reading materials, was fine, as long as "intelligently pursued," the *Post* concluded.[25] In the first great age of the public library and an "earth shrunk by speed," we were already beginning to wrestle with the ramifications of trading depth for breadth. This era marked the first recorded use of the word *literacy*.[26]

Now we dance on the surface of a thousand texts, skimming over billions of words in books and magazines, myriad flashing ads, and across the mesmerizing Web. The text is less the sacred keeper of the flame of knowledge, fraught with meaning and deeply embedded in our psyches, than the transparent carryall of burgeoning info-bits. It's almost an oddity to reread. From the reading revolution emerged the Information Age, which deals in a cognitive currency of "little atoms of content—prepositions, sentences, bits, infons, *morceaux*—each independently detachable, manipulable and tabulable," writes linguist Geoffrey Nunberg. "Information is essentially corpuscular, like sand or succotash."[27] Information is abstract, transferable—your datapoint for mine—and eminently quantifiable, adds Nunberg. Data begin to outstrip the making of meaning, in part because our technologies create an unending stream of information that reveals new arenas of human ignorance, from genetics to the workings of the cosmos, observes Daniel Boorstin. The *Voyager 2* spacecraft, for example, reached Neptune five minutes ahead of schedule on August 25, 1989, after a 12-year, 4.4-billion-mile flight that produced 5 trillion bits of data. This is the age of "negative discovery," a time when we cannot know with certainty "what is" but can only chip away endlessly at what the world isn't, raising more and more questions as we go, writes Boorstin.[28] While he refers specifically to the scientific world, his point holds true for other arenas of intellectual pursuit. We are "captives of information," concludes Walter Ong, "for uninterrupted information can create an information chaos and, indeed, has done so, and quite clearly will always do so."[29] Now we don't ingest the text, we battle to keep it from swallowing us up.

Norbert Elliot doesn't simply teach. He prods, pushes, and spoon-feeds his students. On the afternoon that I visited one of his Documentary Studies classes at the New Jersey Institute of Technology in Newark, a half-dozen students were gathered in a computer lab to present PowerPoints outlining their final projects.[30] The aim was to create a multimedia presentation, and the topics of their choosing ranged from a report on a new video game to a study of the senior experience at the four-year research university. As the students took turns outlining their plans, Elliot repeatedly pounced, interrupting to question their work, especially the validity of their sources. He is a fifty-something English professor in a tweed jacket, with a commanding but warm voice. Like a war-weary general who tries his best to curb his frustrations with the greenhorns, he liberally doles out verbal pats on the back, yet never lets them off easy. "Is that what you're going to use as a standard biography?" he asked Anthony, a junior who was documenting a college revival of a play about New York Mayor Fiorella LaGuardia. (Earlier, Elliot told me that Anthony initially hadn't planned to use any biography of LaGuardia in his research. For his part, the student confided that he'd never visited the school library and hadn't known about its electronic databases until this year.) "On everybody's reviews on Amazon.com, I think it's pretty reputable," Anthony nervously offered, his face glistening with sweat. Elliot suggested that he check reviews of the book on an academic database. "Whatever will be on Amazon would be you or me posting," said Elliot. Most of the other students paid no attention. Heads down, they pecked at their laptops or the lab computers, checking e-mail, and revising their presentations. Soldiering on, the professor urged another student to file for an interlibrary loan book "like today," pushed the video gamer to move beyond online annual reports for an assessment of Nintendo's financial health, and praised a young woman for citing a journal article that she found by accident on Google. "I'm really really going to finish reading that one," gushed the fast-talking Jasmine, who promised lots of flash and photos in her project on seniors. "I got this idea in my head. Hopefully it will come together," she said, almost to herself. "Well done," said Elliot, as Jasmine disappeared behind her laptop. "I think it's going to be good."

Outside class, however, Elliot is often discouraged about his students' ability to sift, use, and make meaning out of the information around them, in other words to "read" their world. He's been the campus gladiator for "information literacy," a buzzword that grew out of educators' realizations in the late 1980s that burgeoning info-gluts were being accompanied by alarming shortages of sense-making, especially among the most tech-savvy of citizens, the young. As the movement took hold, the details began to emerge. Like many of us, students have abandoned print for the Web. As early as 1992, a quarter of library users at the University of North Carolina said they wouldn't use print resources under any circumstances.[31] Shifting attention to the vast resources of the Web sounds natural, and yet faced with a tsunami of largely unsifted, unedited information, students often don't know how to separate the wheat from the chaff nor how to thoughtfully use what they harvest. They—again, like so many of us—are often stuck on the veneer of the information world, grasping at the first "answers" that pop up from narrowly tuned search engines that in turn reveal no more than 15 percent of the highly commercial, poor-quality "surface Web." Even within this tiniest tip of the information iceberg, the choicest bits—relatively speaking—are hidden because search engines such as Google run on algorithmic formulas that place irrelevant or mediocre sites on a par with expert ones. What's highly linked, or paid for, gets top billing, regardless of merit. Yet what comes up first seems fine to most of us: nearly 75 percent of Americans with five or more years of online experience say search engines are "fair and unbiased." Sixty percent don't know that there are paid and unpaid results.[32]

Geoffrey Nunberg was shocked to see that the "highly sophisticated" grad students he was teaching at Berkeley's School of Information Management and Systems had difficulty evaluating simple, unfamiliar information on the Web.[33] A librarian converted English professor Lorie Roth to the cause by showing her printouts of student searches culled from the garbage. "As I sat there, surrounded by the detritus from the trash cans, paging through these print-outs, I tried to reconstruct what kind of thought processes these students were using," says Roth, now vice chancellor of the massive California State University system. "I finally arrived at the conclusion, in fact, there was no process; that there was no logical, clear, systematic inquiry." The realization marked the most disheartening point of her career, says Roth.[34] Study after study at top and lesser schools reveal similar conclusions: students—including seniors—"display a particularly narrow field of vision" in searching, use "quick and dirty" ways of finishing the task, "often opt for convenience over quality," and give up easily. To find the borders of the former Yugoslavia, 35 percent of four hundred and fifty undergrads at UCLA said they'd consult the *Atlas of American History* or the *Encyclopedia of Associations*.[35]

The clincher, and perhaps the best evidence of mounting national concern, is the seventy-five-minute "iSkills" assessment test created in 2006 by the Educational Testing Service, a nonprofit that develops and administers college placement and other exams. The first 6,300 students who took the test scored a mediocre 500 on a scale of 400 to 700. Just half of the test takers could judge the objectivity of a site, and 44 percent could identify the best statement summing up a research assignment.[36] After three years of efforts by NJIT's Humanities Department to push information literacy into its departmental curriculum, seniors who took the first ETS assessment test scored in the sixty-first percentile and freshmen in the twenty-eighth percentile—"a lukewarm performance at best," says Elliot, a reviewer for the exam when it was under development. In a separate internal assessment of information literacy, the writing portfolios of a representative sample of NJIT seniors scored an average 6.14 on a scale of 12, or just below "satisfactory." "Students could find and cite sources better than they were able to judge their relevance and authority, and were even less able to use information they gathered to support their arguments," Elliot and colleagues observed.[37] A tech-savvy expert on educational assessment, Elliot is inured to the "sky is falling" tone of much school reform, and yet this time he's deeply concerned, at times even frightened, by the "lack of authentic engagement," or going deeply, on campus. "If you want to have an educated citizenry, you've got to wrestle with complex ideas," said Elliot, "or you will end up with people who will only do the shallowest things." College is usually our last chance to inspire students to wrestle with depth, context, and problem solving, said Elliot. "If you're going to do it, for god's sake, you're going to have to do it here!"

Definitions of literacy shift over time. Many cultures have counted someone as literate if he simply could sign his name. One Egyptian "scribe" in Greco-Roman times could barely do this, as we can see from a papyrus where he practiced writing his signature, often incorrectly.[38] By the eighteenth century, 90 percent of Swedes could read due to Lutheran church mandates that all should know the Bible, yet few could write. Now, we're setting much higher standards for twenty-first-century literacy by any measure, but at least for now, few are meeting these goals. And what is perhaps most disturbing is that so many are convinced that they are doing well, hunting and pecking through info-gluts on their own. At the University of California at Berkeley, only 7 percent of political science and sociology majors rated their overall information literacy skills as poor, yet 80 percent received poor or failing grades on one comprehensive test. Three out of four students say they are completely successful at finding the information they need on the Web, a level of unwarranted satisfaction matched by the general population.[39] Tech-fluent people aged eighteen to twenty-nine are more confident in their abilities yet less discerning evaluators of online information than other age groups.[40] Throughout history, humans have overestimated their abilities and their wisdom. But hubris is risky. "The greatest menace to progress is not ignorance," says Boorstin, "but the illusion of knowledge."[41]

After class, I hung out in the computer lab and asked Elliot's students how they did their research. Most rarely, if ever, visited the library in person, which isn't surprising given the plethora of scholarly databases available online. Yet only about half of their research was spent culling vetted databases. A good deal of the time, they simply surfed the wider Web, trawling the enormous, churning online information sea. "I think all the information is right at my fingertips," said Jasmine, a solidly built senior with braids and a quick smile. "If I have time, I can search and find anything I want. It's not always the most reliable sources at times, but a lot of cited real documentation is on the Internet. It's like that journal from Duke. All I did was type around and Googled, and it came up." She last visited the library during sophomore year. "I don't go there anymore. I can find it on my own."

We are not born to read. Spoken language is a natural behavior, as evidenced by the evolution of numerous vocal organs from tongue and lips to mouth and throat, notes linguist David Crystal in *How Language Works*. "Unless there is something wrong with the child or something lacking in the child's environment, speech will emerge towards the end of the first year and develop steadily thereafter," he writes.[42] In contrast, there are no organs that have adapted to enable us to process writing, an infant technology roughly five thousand years old. "Reading and writing . . . have to be taught and painstakingly learned," which is why more attention is paid to reading in the long years of schooling than to any other subject. And yet reading, the key to entering the realm of print and its vast storehouses of world memory, long has been a mystery. Written words have conquered the earth, but until recently we've known little of how people decipher them. The eyes swiftly cross a page in little jerky motions that scientists call saccades, after the French for "flick of a sail," then grow still, fixing upon one word per quarter-second, and mystically drinking in its look, sound, and meaning. "In fact, very little actually seems to happen, apart from the eye movement—and these do not begin to explain how the reader is managing to draw meaning out of the graphic symbols," observes Crystal. Nothing much happens if you're on the outside looking in, because learning to read requires a miraculous feat of brain circuitry, fueled by a crucial and equally invisible ingredient: attention. "In asking kids to read, we're asking, 'could you please rewire your brain?'" says neuroscientist Bruce McCandliss. "They have a nice visual system and a language system and are able to make sense of the world. Then we ask them to reorganize their brain so that when they look at a particular visual stimulus, they get speech sounds and the meanings they're linked to."[43]

A boyish redhead with a perpetually eager air, McCandliss works at the forefront of efforts to understand the neuroscience of reading. As a scientist, he says he likes to "make a big mess and see something in the mess that I can focus on"—a characteristically modest way of explaining the breadth of his research. McCandliss studies how people learn artificial languages, uses magnetic resonance imaging (MRI) scanners to peer into children's brains, and tracks the three-dimensional diffusion of water through the brain to measure the efficiency of its pathways—all to piece together how children learn to read. The work is especially pioneering because neuroscientists at first almost exclusively focused on adults, both brain-damaged and normal, to map the cognitive activity of readers, a method akin to "hearing a joke backwards," says McCandliss, associate director of the Sackler Institute for Developmental Psychobiology in New York. Instead of focusing on the "punch line" of how a reader's brain works, McCandliss is endeavoring to uncover the full story of how our brains are molded into reading machines, from the "opening line" of how language is heard and words are seen to the meaty middle of how readers move toward literacy. If reading is an invention, a technology to which we adapt in order to draw meaning from marks on a page, then understanding how people learn to read can give us clues to how experience shapes the brain. And if building a literate brain involves the slow creation of a cognitive scaffolding, then understanding this process may teach us how to build structures of understanding in an info-saturated time. The essence of reading, after all, is going beyond the surface.

How do we read? The eye is the gateway, the first connection between man and word. Yet sound, ironically, is the foundation of reading, as perhaps is fitting for a species that developed complex oral communications long before writing was born. A child must be able to understand and manipulate the sounds of language, from syllables all the way down to *phonemes*, the smallest units of sounds in speech, before she can read. This skill is so crucial that a low capacity to discriminate phonemes in *infants* links to later difficulties in learning to read, and sheer phonological skill in kindergarten predicts teenage reading ability better than whether or not the five-year-old could read upon entering school.[44] Kids who can't hear the difference between "ba" and "pa," for instance, may be at risk. Dyslexics, in essence, suffer from a kind of cognitive "deafness," in that they are not as sensitive to letter sounds within words as skilled readers are. They don't recruit the brain's left perisylvan region—the area associated with phonological processes—and so are trying to read with a much-diminished sense of the language. Asked to do a silent, written rhyming task, skilled adult readers—but not dyslexics—will show heightened activity in the perisylvan regions that deal with auditory comprehension. We really do have a voice in our head that brings the written word to life, making music with the silent text.

At the same time, we couple this wondrous ability to "see" sound with a unique visual expertise—the recognition and comprehension of words. As children learn to read, they are honing an area in the brain— the "visual word form area" in the left fusiform gyrus—to respond quickly to written words. This tiny area, related to seeing complex visual shapes, learns to respond far more to words, or even letter combinations that closely resemble words, such as *blard*, than to spoken words or groups of nonsensical letters, such as "*Rbldf*."[45] This process is long and difficult, lasting well into adolescence and perhaps beyond.[46] But the development of the word form area is so necessary to reading that patients who suffer lesions in the left occipital and temporal lobes—that house the visual word form area—lose their ability to read whole words, although they can sound them out letter by letter. In essence, they read like beginners again. English speakers apparently develop particularly robust visual word form activity in the area of the brain—the left ventral occipital lobe—related to recognizing complex shapes because the language's irregularities demand careful circumspection. (*Pint* and *lint* sound far different than they look, forcing a reader's brain to work hard to integrate their graphic and phonological characteristics.) In contrast, Italians digest words in areas that snuggle up more closely to sites of auditory comprehension, since the sight and sound of their language deviate less often.[47] For most languages, fluent reading most often involves a strengthening of visual capabilities on the brain's *left* side, the hemisphere that is home to our powers of speech.

It's not enough, however, that these crucial skills of recognizing words and knowing how they sound take root in neighboring areas of cognition. The visual and phonological areas in question must be strongly linked in what McCandliss describes as nothing short of a "reorganization" of the brain.[48] While neuro-circuitry is surprisingly changeable and even a single thought can create a novel connection between brain areas or even between neurons within an area, many aspects of cognition—tying your shoe, using a fork—become reflexive, with a well-worn brain circuitry. For a child to become a fluent reader, the elemental work of reading must be automatic in this way, or he will never be able to explore a text's meaning. He will be stuck on the surface, wrestling with simply recognizing each word, and missing the tapestry of their cohesive whole. Moreover, the key to this reorganization is *attention*. Focusing on the most basic of letter-sound connections in a word—"*a-t*" sounds like "*at*," for example—literally carves out neural pathways between the visual and phonological areas involved in reading. In pilot studies in New York classrooms, McCandliss has taken eight- to twelve-year-old poor readers and bumped them up one or two grade levels in reading ability in just twenty sessions of forty minutes apiece by practicing the most simple letter-sound combinations again and again. Local teachers, burned by the reading wars and initially opposed to his seeming "drill and kill" methods, aren't as skeptical of his techniques anymore. The work isn't merely teaching the decoding that is now known to be essential to learning to read but also harnessing the power of attention to drill this activity into the brain.

To illustrate, McCandliss showed me the latest artificial language he has taught to adults in order to test how the spotlight of attention can drive learning. On his laptop screen appeared a curvaceous character that looked like a decorative bit from an illuminated manuscript, with the sinewy top third representing "c," the middle third "a," and the bottom bit meaning "t"—in total, "cat." It's as if the letters c-a-t were fused together, not just lined up alongside one another as they are in a word. In this made-up language, the letters and word-characters look nothing like our alphabet, so learning the symbol for cat or dog is no easy task. In one experiment, sixteen people were taught to focus on learning the whole word-symbols, such as cat, dog, or table, while another sixteen were told to pay attention to the individual letters within each symbol, including their letter-sound connections. Those who focused attention on the letter-sound connections that form the characters remembered the words longer and were better at learning new words. They also showed more activity in the left hemisphere of their brains, while those who focused on learning whole-words from the start exhibit a diffused pattern of bilateral cognitive activity— much like beginning readers who haven't mastered left-dominated visual-verbal systems of reading. The letter-sound learners essentially were etching more efficient reading pathways in their brain, much like the New York schoolkids. "The punchline is that the way you focus your attention during learning has a profound impact on the brain's response," explained McCandliss, sitting up, his thoughts tumbling out quicker and quicker in his excitement. "Attention is like the needle that stitches new patterns in the brain!"

Perhaps this is why those on the cusp of the eighteenth-century reading revolution, who still experienced reading as an intense, visceral endeavor, commented so often on the sheer physicality of the process. They may have intuitively understood what our technology permits us to newly discover, that reading is a demanding activity that shapes you in mind and body. "Eighteenth-century readers attempted to 'digest' books, to absorb them in their whole being, body and soul," writes Robert Darnton. "The physicality of the process sometimes shows on the pages. The books in Samuel Johnson's library . . . are bent and battered, as if Johnson had wrestled his way through them."[49] A century earlier, influential surgeon Thomas Willis theorized that reading involved channeling visual imagery into the brain via "animal spirits" that moved through the body and mind but, like emotion, needed to be restrained.[50] Good literary habits could keep man's wild side in check, and yet an excess of reading or consuming the wrong books could be dangerous. "Susceptibility to colds, headaches, weakening of the eyes, heat rashes, gout, arthritis" were among the ailments attributed to too much reading by J. G. Heinzemann in 1795 at the dawn of the "reading revolution."[51] Now we still worry about excesses, but we don't believe that reading will make us physically sick. Except by a very few, reading is not seen as sacred, intensely challenging, or even as the "search for a difficult pleasure" that Harold Bloom suggests.[52] Today, we more often skate across prose—be it on screen or printed page. We read to glean and to get the neatly packaged answer, not to be changed. We read to gather information, says Carol Collier Kuhlthau, middle school librarian turned world-renowned sleuth of the "information search process."

Surely, this isn't a problem. Retrieving the choicest, most valid bits of text from the oceans of information around us would seem to be the sine qua non of sense-making. Wrong, says Kuhlthau, a Rutgers University professor and the closest thing to a celebrity you'll find in the library world. That's only part of the story, she asserts. More than two decades ago, Kuhlthau did for information seeking what McCandliss is doing for reading: she uncovered a process of the mind that had remained a mystery for hundreds of years.[53] Beginning with her own students and continuing with hundreds of adults, she discovered how people search for information, from their first vague desire or need to know something to their concluding paper, presentation, or quitting point. While philosopher John Dewey, psychologist Jerome Bruner, and others had grappled with the links between knowledge and learning, and librarians had studied whether users wound up with the "right answer" or "right source," Kuhlthau was one of the first to investigate the wider context of information seeking—the dynamic interplay of mind, body, and spirit that occurs during sense-making. What she found was a librarian's nightmare: a messy, painful, emotional process that, like reading, takes great effort and attention to do well. She also found that most people, from workers to kids to scholars, don't go deeply enough into the search process to do much more than collect a grab-bag of information-morsels. They are left with, to paraphrase Dewey, a "cessation" of their task, not the "consummation" of their quest.[54]

How do we search? How do we make meaning from information? Whether we're looking for information on a diagnosis of cancer or writing a work report, uncertainty is an essential component of the process, not an unwelcome impediment to learning. In her five-step roadmap of information searching, Kuhlthau found that people experience apprehension as they approach a research task and feel confused and unconfident as they focus in on a topic or area of interest. In the third and most difficult stage, "exploration," our uncertainty, doubt, and confusion increase as we dig into the dispiritedly inconsistent information pertaining to the subject. "When a person is involved in the dynamic process of becoming informed, relevance does not remain static," Kuhlthau says dryly.[55] At this point, we are poised to construct meaning if we can tolerate and push past the ambiguity and discomfort impeding our journey. The next step is most crucial: finding *focus*, the framework for the quest. "As the saying goes, a question well-put is half-answered," wrote Dewey. "In fact, we know what the problem exactly is simultaneously with finding a way out and

getting it resolved."[56] With focus, interest in the proje͞
dence builds, evidence is collected, and the long, hard tasҝ
mated. We can see why David Crystal calls written language
ative process, an act of discovery," and Bloom cautions that "one ;
be an inventor to read well."[57] Reading literally and in a larger sense
is an act of will, spirit, attention, and confidence. Like Johnson, we
must wrestle with our texts.

And yet few do. By the end of a search, only a quarter of people
are collecting evidence to bolster a focused topic, and half never attain
a focused perspective on the topic *at anytime* during the process,
according to Kuhlthau's research.[58] We in the digital age didn't invent
the missed opportunity, as Dewey reminds us. "There is distraction
and dispersion, what we observe and what we think, what we desire
and what we get, are at odds with each other," he wrote in 1934. "We
put our hands to the plow and turn back; we start and then we stop, not
because the experience has reached the end for the sake of which it
was initiated but because of extraneous interruptions or inner
lethargy."[59] The possibility of clinging to the surface of life has always
existed. And yet in honing the art of skimming across infinite texts are
we losing the hard-fought skill of delving beneath the surface? "To
find something valid (on the web), you really have to dig through it,"
said Prashant, one of Norbert Elliot's students. "But there's so much
that you never really have time to dig through it. You end up digging
through so much useless stuff that you get tired of looking and you end
up looking somewhere else, and you never find what you started to
look for." Anthony, the student studying LaGuardia, described his
style of searching more succinctly. "It's just get what you can and
move on." Jerome Bruner stressed man's capability to actively make
sense of the world. George Kelly described how people shift and
amend their "constructs," or frameworks for understanding: faced
with new information, they either grapple with its implications, or turn
away, clinging rigidly to old, unchanging readings of the world. Faced
with the challenge of sense-making, are we opting for the passive
way? Are we losing the will to plunge again and again into the infor-

mation muck to retrieve the rough gems of meaning? Are we keeping our hands clean?

In the end, literacy is what we make of it, and unless we want to slowly redefine it as a business of skimming, we must find new ways to dig beyond the surface of the text and into its depths. To start, we should recognize the exertion that all effective reading demands. Creative reading—whether of book or screen or scroll, or more widely, the world around us—is a *building* toward knowledge. Even the accessible, seductive texts of cyberspace will not spoon-feed us. To navigate the literate world, we must build a literate brain, then work past uncertainty to construct our own frameworks of interpretation. But we cannot do all this alone. The *context* of our sense-making is as important as the form of the text and the force of our will. Throughout history, each age has produced structures of discourse and understanding—from rhetoric to the "order of the book"—to cradle the voice and then the text, and form the foundation of man's approach to their message. For we cannot make meaning out of the unsifted and the random, no matter how hard we try.

Ancient peoples balanced the fluid immateriality of the oral with the secure structures of place, rhetoric, and memory. "For the Greeks, who assiduously wrote down their philosophical treatises, plays, poems, letters, speeches, and commercial transactions, and yet regarded the written word merely as a mnemonic aid, the book was an adjunct to civilized life, never its core," explains Manguel. "For this reason, the material representation of Greek civilization was in space, in the stones of their cities."[60] Mirroring this built order were architectures of thought: rhetoric, and especially memory. Classical rhetoric became the core of a good education; its mastery the mark of a successful man. Its five disciplines—*invention* or focus, *arrangement* or organization, *style* or diction, *memory*, and *delivery*—strongly influenced early literature, according to classics scholar George Kennedy.[61] Yet long before rhetoric, how could people create and preserve complex thought? "The only answer is: Think memorable thoughts," says Walter Ong. "Your thought must come into being in heavily rhythmic, balanced patterns, in repetitions or antitheses, in alliterations and assonances, in epithetic and other formulary expressions."[62] Proverbs such as "Red in the morning, the sailor's warning; red in the night, the sailor's delight" were not linguistic gloss but rather formed "the substance of thought itself," says Ong. Such concepts are nearly impossible to appreciate today, when we outsource memory almost entirely to our gadgets. In the late 1920s, Milman Parry shook the linguistic world by demonstrating that Homer's *Iliad* and *Odyssey* were oral creations, since nearly every line was formulaic. The phrases *metephē polymētis Odysseus* (there spoke up clever Odysseus) or *prosephē polymētis Odysseus* (there spoke out clever Odysseus) occur seventy-two times in the poems. A bard in an oral culture does not memorize a poem, as we would a text, but rather restitches the work together in performances that are never the same twice. Hearing a tale, a bard often will wait a day before retelling it, in order to let it "sink into his own store of themes and formulas," notes Ong.[63] Learning to read and write ruins an oral poet. Unavoidably envisioning the poem as a text, a literate bard no longer can freely stitch the old work anew from his own stock of story threads. Once upon a time, stories grew out of fabric of our spirit. We were one with them. Along came information: present-bound, unambiguous, surface. This is why information is a menace to storytelling, observes Walter Benjamin, and why storytelling is slowly coming to an end.[64]

It took one thousand years to create for the book what the architectures of place, rhetoric, and memory did for the spoken word. The book will always change with us in both form and in content, but alongside its mutability, a complex culture of print has developed that effectively structures written thought. We can be fairly certain that a book we purchase today is written by the named author and produced by a reputable publisher with the author's consent, and we can be sure that this book is the same as one bought in London or even, in translation, in Munich, asserts Adrian Johns.[65] At the time the Morgan Library's *dos-à-dos* Bible was born, none of these assumptions could hold, as Johns details in his cultural history of the early age of print, *The Nature of the Book*. Fixity and veracity were hard-won cultural artifacts forged over hundreds of years. The main feature of the early print age was "uncertainty," writes Johns, until systems of copyright, editing, publishing, cataloging, classification, licensing, and distribution were hammered out, especially through the efforts of London's Worshipful Company of Stationers, the printing trade's guild. If writing stilled the spoken word and made language a visible and external thing, and print's great outpourings left us floating on the surface of a sea of text, nevertheless the order of the book hammered out safe pathways *back in* to the meaning.

Now, we seem to be hurrying past these hard-won achievements in our rush to embrace what Marinetti called the "free and random" word. Contrast the great French Enlightenment *Encyclopédie* with the largest, most radical reference work of our time, the online resource Wikipedia. Both began as subversive attempts to disperse knowledge to the masses and were enormously successful in this mission. The first of the French compilation's twenty-eight volumes sold out quickly and continued to be a best seller despite a steep price and a brief church-instigated royal ban.[66] The nonprofit Wikipedia gets as many as fourteen thousand hits a second and is the seventeenth most popular site on the Internet, according to historian Stacy Schiff.[67] Yet while founder Jimmy Wales calls himself an "Enlightenment kind of guy," his rationale in founding Wikipedia differs greatly from the French work's authors, Denis Diderot and Jean le Rond D'Alembert. "The goal of the *Encyclopédie*," wrote Diderot, "is to assemble the knowledge scattered over the surface of the globe and to expose its general system to the men who come after us, so that the labors of centuries past do not prove useless to the centuries to come."[68] Diderot sought to create a "library" that would make visible the connectivity of the "tree of knowledge" through energetic cross-referencing. (It was this aspect of the work that caused trouble for Diderot, who tweaked the church by, for example, referencing the Eucharist to cannibalism.)[69] In contrast, Wales set out to combat ignorance through offering a massive—one million articles and counting—and free compilation of any and all information written by anybody with a keyboard. We now have instant access to articles on the green-blooded New Guinea skink, the developer of high-fructose corn syrup, a series of 1981 sewer explosions in Kentucky, and the Scottish railway station of Jordanhill, the million-mark entry that, according to Schiff, was edited more than four hundred times in the first day of its posting. Is this the long-overdue realization of collective knowledge that Pierre Teilhard de Chardin called the "noosphere" or just a hodgepodge microcosm of the often-inaccurate Web itself?

For all Wikipedia's unparalleled popular success, the encyclopedia gets so-so marks for accuracy and, more importantly, for vetting and ordering information in a way that promotes knowledge creation. Entries are supposed to have a neutral point of view, with verifiable content that's been previously published, yet they are often prey to vitriolic edit wars that hardly further the cause of truth. In response, Wales has appointed administrators to police the site, set up arbitration and mediation committees, and placed a ban on anyone's ability to reinstate an edit more than three times in twenty-four hours. Pages dealing with coordination and administration are proliferating. They made up 85 percent of the site in mid-2006, compared with 70 percent nine months earlier, writes Schiff. Even so, "it can still seem as though the user who spends the most time on the site or yells the loudest wins," she writes, noting, as well, that Wikipedia is largely written by a tiny slice—2 percent—of its two hundred thousand registered users. A "voice for all," at least in theory, is a good way to run a country—it's called democracy—but it may not be the best means to create and disperse *knowledge*. Much squabbling accompanied the sixteenth-century birth pangs of the "order of the book," and yet the literary midwives of this time, for all their particular elitism, endeavored to create systems for promoting accountability in print, not spaces where one man's tirade could masquerade as, and be exhibited proudly alongside, an established truth.

What Wales essentially wants to do is to create a kind of mini-version of the universal library, the repository of all knowledge that has long been a dream of literate peoples. The digitization of major research libraries marks the largest recent effort to begin building a universal library that would be "truly democratic, offering every book to every person," enthuses Kevin Kelly. "Once digitized, books can be unraveled into single pages or be reduced further, into snippets of a page. These snippets will be remixed into reordered books and virtual bookshelves." With the linking of all parts of each digital text, says Kelly, "the universal library becomes one very, very, very large single text: the world's only book."[70] The image is astonishing: the book is unbound, its pages scattered, and prose disconnected. Here is a context of reading we cannot ignore. For the hypertextual Web takes information—the detachable, interchangeable, flattened "atoms of content"—and makes it boundaryless. "This is a question of geography rather than structure," says Geoffrey Nunberg. "It's significant that infospace is rarely depicted by comparison to anything more structured, like a city."[71] The goal—all knowledge, all the time— seems worthy, but the reality can be, in John Updike's description, "grisly."[72] Or perhaps dizzying is a better word, for even small brushes with hypertext leave many people reeling. In one British study, those exposed to hypertext set up in a linear manner found navigating a site easier, more memorable, and less disorienting than working with typically more intensely linked sites. Differences in learning styles matter. "Field independent" learners, who are more individualistic and less influenced by others, jump around sites and do better at information searching in this milieu. In contrast, holistic and more social thinkers prefer linear Web worlds. But in general, disorientation is common in hypertext, studies show.[73] The universal library ultimately would give us not knowledge but a pale, virtual echo of the real world, an infinite *inscripto continua* that we might or might not be able to decode. In the Jorge Luis Borges story "The Congress," an Uruguayan rancher tries in vain to create a complete encyclopedia of all knowledge. On the night he gives up, he wanders the city, seeing people, houses, the river, and a marketplace. Finally, "he realizes that his project was not impossible but merely redundant," writes critic Alberto Manguel. "The world encyclopedia, the universal library, exists, and is the world itself."[74]

In the end, perhaps we are asking too much too quickly of our newest tools. Scattershot efforts to teach information literacy have been mounted on campuses around the country, but most are Web-based updates of tired bibliographic instruction classes that shuffle freshmen in and out of the library in fifty minutes, few questions asked. A growing number of brave souls, including Elliot, have worked with faculty to embed the library world's standards—define, access, evaluate, manage, integrate, create, and communicate—into the curriculum. Still, years of efforts, impeded by campus turf wars and terminology spats, have yielded minimal progress, as the new ETS assessment shows. "For over 20 years, campus projects designed to encourage students to develop a sophisticated understanding of the library and to develop their information skills have met with minimal success," report E. Gordon Gee, president of Ohio State University, and Patricia Senn Breivik, a retired university librarian who spearheaded the movement.[75] Surely, the discipline is defined too broadly: it's supposed to teach technological prowess, critical thinking, and research skills, prompting Fairleigh Dickinson University's chief librarian James Marcum to quip that promoters want to turn "every information seeker into a librarian,"[76] a perhaps worthy but impossible goal. More importantly, the movement largely transfers the battleground of literacy to the unbounded realm of cyberspace, using the very tools that tend to keep us on the surface of the text. We're often asking information seekers to become not just librarians but to re-create for themselves the thousand-year achievements of the age of print.

Before I left Newark, Norbert Elliot told me a story. To great fanfare in the local media, he recently stopped lecturing and instead began requiring students to listen to podcasts of his lectures before class, then come prepared to discuss the material in depth. The trouble is that Elliot suspects that the few students who do listen to the podcasts are usually multitasking or running about while doing so. They probably are not listening carefully or taking notes, as at least some might in class. "We had three news channels cover our release on iTunesU, and we all sat there and said what a great thing it is to have, and on the other hand, while I'm saying that, I'm thinking to myself, 'But I know my own students, and just because they have this wonderful technology, it's probably seduced them in some way to think that they can do this while they're doing their math homework or something," recalled Elliot. "I have sort of sold them a questionable bill of goods about how easy this technology is to use." At the same time, Elliot is careful not to push his students too much, for when he does, he loses them. He and other professors who cajole and demand students to dig deeply tend to have high class withdrawal rates and poor scores on teacher-rating sites. Elliot, who's been dubbed a "stalker" on one site for pursuing students through every step of assignments, obsesses over how best to inspire students without scaring them off. "There's real resistance when we ask students to do these things," said Elliot dispiritedly. "They go away, because they can go someplace where someone is not asking that of them." They go where the "good enough" answer suffices.

Our new tools are built for movement, for gliding freely across worlds of info-bits floating in an everlasting present. Surely, we should prize skimming as an increasingly useful skill in a time of information overload, just as we can appreciate the screen as a labyrinth of new textual and visual experiences. In the end, Norbert's student, Jasmine, produced an evocative multimedia presentation on the bittersweet experience of being a college senior that combined a three-page essay with a colorful slideshow of campus snapshots. And perhaps her project's title, "NJIT Seniors: Going Forward, Looking Back," evokes, as well, our own nascent efforts to balance on the cusp of a new age. But now is the moment to ask ourselves: do we want to build a culture that relies predominantly on skimming? To drift steadily on and on, distracted, across texts, is as much like deep reading, as stockpiling information is akin to acquiring knowledge. To fully understand the rich intricacy of any writing—from the multilayered symbolism of a novel to the nuanced arguments of great nonfiction, we must go *deeply* into the text. This is an unavoidably laborious, often uncomfortable yet inevitably rewarding process.

So for now, it is too risky to prematurely detach the art of reading from the realm of the book. "In the electronic anthill," asks Updike, "where are the edges?"[77] The mutable, sociable ever-shifting book, which brought us to the doorstep of the digital age, must be celebrated for its *edges* and as the product of hard-won structures of fixity and accountability. Eventually, we may create an "order of the screen," although, as Nunberg rightly notes, the porous Web resists the solutions imposed by the era of print.[78] Something of a nascent effort can be found in the messy middle ground where book meets screen, with the "networked book." One such work-in-progress manuscript, a book on video game theory by McKenzie Wark, was posted online for reader comment before its eventual publication both online and by a university press.[79] A wiki-book? No, the experiment by the Institute for the Future of the Book and Wark, a professor of media studies at the New School University in New York, tuned up the interactivity of creation a notch, yet Wark retained authority over a text validated by a traditional publishing process. Is this then a case of a networked book, or a bookish screen? No matter, for even if the book becomes looser and the screen more orderly, we'd be wise to make good use of both and value each for their strengths as we go forward, while also remembering that in either realm, we urgently need to value the wisdom of gatekeepers such as Norbert Elliot, Carol Kuhlthau, and John Bidwell. They insistently lead us into the depths of reading, pulling us toward the light.

Carefully, Bidwell boxed up the *dos-à-dos* Bible before its return to the library's vast vaults dug three stories down into the bedrock of Manhattan. Still and silent, the relic yet offered a parting message, a reminder that books provide an unending adventure not only of the mind but of the senses. Many books have a distinct and often beloved olfactory signature; Bidwell can still smell on some the aroma of the cigars that certain antiquarian booksellers smoked long ago in their shops. Reading, as well, transforms mere marks on a page into meaning through a powerful blending of sight and sound. And lastly, a book neatly links vision and touch, producing a perfect symmetry of distanced and proximate discovery. "To know something as fully as possible, we need to be close to it . . . , and we need at the same time to be distanced from it, to have it 'in perspective,' an object notably distinct from ourselves," writes Walter Ong.[80] In an age of virtual, mobile, split-focus distraction, the book is a link both to body and spirit. To read a book is a grounding and an ascension all at one moment, a feat no computer can yet carry off.

Chapter Seven

AWARENESS

The Post-Human Age:
A Battle for Our Attention

Aaron Edsinger was trying to imagine life without his robot. He designed and built Domo, one of the world's most advanced robots, from scratch for his doctorate at MIT's Computer Science and Artificial Intelligence Laboratory. Over three years, it grew from a pile of motors and springs and circuit boards into a coarse, eyeless, legless mechanical man, and then, after countless days-stretching-into-nights of work, into Domo—a touchable, teachable, blue-eyed robot whose strength of presence has startled even Edsinger. "I think what really surprised me working with it for three years is the feeling that this thing developed from being an object to something more like a creature, at least in my relationship to it," observed Edsinger, the Seattle-born son of an aerospace mathematician and part-time cattle farmer.[1] Edsinger spent years living on the cheap, sleeping in his car for stretches at a time, piecing together junkyard-cool robots for the San Francisco performance art scene before landing almost by accident in one of the world's most respected artificial intelligence labs. Compared to the True Believers who solemnly predict that robots will be smarter than humans by 2020 and that we'll be downloading our brains onto computers, Edsinger is practically a skeptic. He's not a geek, doesn't read science fiction, and doesn't relish the "more technology, the better" mantra of the trade. He's more like an inventor-philosopher, in high tops and jeans. And so when Edsinger talks about Domo's "presence" and how he vacillates between thinking of this robot as a tool and a *friend*, you can't help but be intrigued. Watch Edsinger play or work with Domo, and you'll get a peek at the complexities of our future relations with intelligent machines. Observe Domo and its creator and you'll see the possibilities and dangers before us as we increasingly share our planet—and our attention—with an unprecedented array of artificial life.

"Hey, Domo." Edsinger greeted the robot to catch its gaze, perhaps for the last time. Done with his graduate studies, Edsinger was about to return to San Francisco, while the Toyota-sponsored Domo had to stay at MIT to become the subject of more student tinkering or perhaps an inductee into the campus museum. On the day of my visit, Domo loomed as a kind of gatekeeper at the entrance to Edsinger's cubicle, which was strewn with packing boxes. Domo, whose name means "thanks" in Japanese, is legless and skinless and fused to a table, yet its metallic intricacies exude a kind of aesthetic of the complex and the precise. Its beauty is industrial, like a skyline or a clockwork. But it moves like a human, reaching, grasping gently with three long fingers covered in a soft, smooth plastic, speaking in an eerie, melodic voice. Unlike factory robots that largely do highly scripted tasks and are dangerous if impeded, Domo is designed to work with people in an everyday space such as a kitchen. With twenty-nine "degrees of freedom" or motorized joints, it can track movements, especially faces, with its huge roving eyes and can gently grasp everyday objects, such as a ball or paper cup, and set them down. Perhaps most importantly, it can sense touch, so its spring-loaded hands and arms can be led. But if you push on it too hard, it languidly and melodically says, "ouch." Their constant physical contact, said Edsinger, is one reason he's thought of Domo as clearly an object and yet a partner, too. "When it grabs my hand, there's a very visceral reaction," said Edsinger, who is compact and wiry, with a gap-toothed smile, warm brown eyes, and a calm, approachable air. "You really feel like there's something there hanging onto you." Moreover, because Domo is built to react to its environment, it does the unexpected, like we do. And that can be alluring. "You know that it's a piece of metal and motors," Edsinger said. "And you know all the code that's going on behind it but every now and then it acts in a way that you don't expect. So sometimes you get that glimpse of it being very life-like."[2] Clearly, Edsinger was going to miss Domo. "I certainly have been thinking about what it will be like to leave it behind and move on and see it appear in some press blurb and be like, 'that's my guy,'" he said. "The thing is, I felt like a sort of social relationship with it comes very naturally." In one photo, a smiling Edsinger stands with his arm protectively around Domo's shoulders, while the robot stares into the camera, mechanical eye to eye.

Gentle Domo is no Frankenstein; it's designed to be part of the family. Entranced, we're setting a place at the table of our lives for sociable machines. Once upon a time, startling eighteenth-century automatons, such as Jacques de Vaucanson's famous flute-playing android or Wolfgang von Kempelen's infamous not-quite mechanical chess-playing Turk, were built not to befriend but for show. They were "philosophical toys," notes Gaby Wood in her quirky history of the automaton, *Edison's Eve*.[3] Centuries later, postwar scientists pinned their hopes on constructing perfectly rational and disembodied, replicated brains that could play chess and crack equations like walnuts. But now a new leading edge of artificial intelligence focuses on the creation of "affective," or emotional computing and the construction of robotic devices that can serve as housekeeper, therapist, receptionist, and coach, and may one day become judge, teacher, and friend to the willing. Roomba, the best-selling robotic vacuum cleaner created by Edsinger's adviser Rodney Brooks, and smart toys like Robosapien, eventually will be footnotes to the story of the rise of the personable machine. In a day not too far off, Domo will set the table, feed the baby, and swap tales of life's "ouches" with us. Hey, Domo—indeed.

At the same time, we are blending with our tools and blurring the boundaries between human and machine in unprecedented ways, through neuro-implants and implanted RFID (radio frequency identification) chips, smart prostheses, and pharmaceuticals that can change mood, memory, or focus in push-button time. Machine becomes manlike and man becomes machinelike, a convergence that may augment our capabilities yet reduce our humanity, however we define it. "No longer is it possible to see the human subject and the machine as aligned but separated entities, each with their own separate functions," writes geographer Nigel Thrift. "No longer is it possible to conceive of the subject as simply 'alive' and the machine as just so much 'dead labor', the two linked only by their capacity for movement."[4] Our burgeoning relationship with the machine, with the "spirit" of the machine, and with our own machine-stamped spirit ultimately risks narrowing our connection to ourselves and to each other. This is the third collective loss, after trust and depth of thinking, that we face in a time of distraction. When we embrace the machine not as a tool but as part of us and as one of us, we begin to lose the inner will and outer means to connect with one another. We risk living in solitary glass cages, enchanted by shadows on the wall.

Just steps across the MIT campus from Domo's home base is a bustling hotbed of work on the future of the sociable machine. Deep within the vast labyrinth of the university's Media Lab are the side-by-side laboratories of professors Cynthia Breazeal and Rosalind Picard, kindred spirits and close collaborators in the effort to create robots and computers that we can relate to. They are both brisk, stylish, and workaholic, female gladiators who endeavor feverishly to soften the edges of fields pioneered by ruthlessly cerebral—and some might say, unsociable—men. You'd never think "geek" when you first meet the dark and cool Breazeal or the fair, wholesome Picard. Their workshops within the Media Lab are cluttered and homey. Breazeal is the roboticist, schooled at the AI laboratory where Domo took shape. Picard is a computer scientist with a background in electrical engineering who tuned the family car as a girl in Georgia. Both seek nothing short of a revolution in our interactions with machines. Instead of a computational monologue, they want a conversation. In place of a task-centric artificial intelligence, they envision a smart friend. In essence, they want to remedy robots and computers of their social deficiencies, cure them of their mechanical autism. "Lean on me, confide in me, carry on the intricate social dance that you do with fellow humans," these scientists effectively want the computer to invite. And all evidence shows that we humans will eagerly accept their offer.

For her doctorate, Breazeal created Kismet, a captivating, elf-eared robotic head that charmed people through infantile babbling and well-timed empathetic coos and movements. When work was done and this revolutionary robot became an MIT museum piece, Breazeal felt a "sharp sense of loss," her colleague, psychologist Sherry Turkle, has written. "But Cynthia is Kismet's mother," Turkle quotes one visiting ten-year-old as lamenting on one of the last days that Breazeal had access to the robot in 2001.[5] Now heading her own laboratory, Breazeal is a self-described "creature builder," a constructor of mechanical "beings" that emerge from the recesses of her lab into the sunlight to guide and enchant us. The robotic teddy bear Huggable, built as a companion for children and the elderly, has two thousand sensors under its fur-covered silicon skin that can measure temperature, force, and electrical fields.[6] Embrace it, and it hugs back. Put it on your lap, and it looks up at you with video camera eyes and nuzzles into your arms. Set it on a table, and it can look around, catch your gaze, and gesture to be picked up. All the while, the robotic bear relays audio, video, and other information wirelessly to a nurse, tutor, babysitter, or parent. Is the Huggable being hit or ignored? Has the child fallen or gotten distracted? The Huggable knows, and can tell. The robot is a form of "local intelligence," which can "reduce the cognitive load of the operator," explains Breazeal. A Huggable can also strengthen family ties, enabling a soldier, business traveler, or faraway grandparent to be with a child virtually. "Tuck Your Child in Bed from Across the World," reads a PowerPoint slide hanging over Breazeal's shoulder as she unveils the Huggable to corporate sponsors. "Robots can engage us like never before," says Breazeal, who confesses to yearning for a Huggable who will tutor her three young sons in a foreign language. "They can really push our buttons."[7]

As well, she and Picard are working together to oversee the creation of RoCo, one of the world's first expressive computers. Based on research showing that the body language of both teachers and students influences learning, RoCo tracks the posture of a user and, like an animated flower, shifts its monitor up and down, forward and back, to either mimic or *influence* the user's stance. Slouch, and the monitor might tilt forward and drop almost to the desk. Straighten up, and the computer swings forward and upward. Alternatively, the computer can subtly direct a user's posture by taking a different position than the user at any one time. Beeps and coos, designed to resemble R2-D2 in *Star Wars*, are planned to boost the machine's expressiveness. In these ways, the computer is to build rapport with the human. Most machines treat people like furniture, ignoring their goals, needs, behavior, and emotions, observe Picard and Breazeal. As a result, they cannot help support or "prioritize" your pleasures and needs. But an expressive computer that is more likable can go a step beyond Microsoft's prototype interruption machine, BusyBody, and become our companion, our workmate, or even our child's schoolmate.[8] Already, sociologist Clifford Nass has shown that computer voices in cars that reflect driver mood—subdued when the driver is upset, energetic when the human feels up—promote safer driving and more attention to the road in driving simulations.[9] Noting that children have been shown to learn more when they do schoolwork together with friends than with nonfriends, Picard and Breazeal suggest that a RoCo would make a good study-mate. "The system would not only acknowledge the presence of the child and show respect for her level of attentiveness, but also show subtle expressions that, in human-human interaction, are believed to help build rapport and liking," write Breazeal and Picard.[10] *Let go, RoCo. I don't want to hold your hand.*

Are these cheap tricks or a much-needed solution to the frustrations of increasingly living side by side with smart yet infuriatingly stupid machines? It's true that we are surrounded by massively complex automated systems that hardly know we're there—from smart buildings and cars to vast telecommunications and data retrieval systems. Automated systems decide what medical test you get, whether you should win a job interview, a loan, or a shot at a particular date on a Web matchmaking site. And rather than an intelligent dialogue with our machines, we talk *at* each other, according to Don Norman, computer design guru. "When the machine intervenes, we have no alternatives except to let it take over: 'it's this or nothing,' they are saying, where 'nothing' is not an option," writes Norman. "We issue commands to the machine and it, in turn, commands us. Two monologues do not make a dialogue."[11] Norman, like Picard and Breazeal, sees affective computing as a way to soften the machine's unbending edges and enable them to take a more workable place alongside humans in society. Today's logic-centric machines are superior in speed, power, and consistency, yet lacking in social skills, creativity, and imagination, says Norman. In other words, he also suggests a less autistic machine, which can "read" and react to us as a horse does to a rider. In effect, this not only bonds the machine to us but boosts its intelligence, since emotion is becoming increasingly recognized by researchers as a crucial component of intelligence. German and American government scientists are building car and airplane systems based on this "H- or Haptic-Mode," or horse-rider theory, so that pilots/drivers can give automated systems more or less control, in other words, tight or loose "reins," depending on the circumstances.[12] Such efforts underscore how our machines are becoming part of our environment, much as the air that we breathe. We no longer have a choice of whether we want a relationship with our machines. What matters now, as Turkle points out, is what *kind* of relationship we will develop with them. Will we ask Domo to make toast or listen to our woes? Is it a tool or a person?

"How are you today? Can I have a kiss?" A stocky woman with flaming red hair is talking with a baby-faced robot named MERTZ created by roboticist Lijin Aryananda, formerly of MIT and now a postdoctoral research associate at the University of Zurich. At a panel discussion on human-machine interaction at the American Museum of Natural History in New York, Aryananda showed video clips of MERTZ, which consists simply of a wide-eyed, expressive face on a swaying, flowerlike stem. Since getting the robot to recognize faces was the raison d'être of her doctorate, Aryananda for years exposed MERTZ to as many "naives"—newcomers to the technology—as possible. Parked in an MIT lobby like a cartoonish ATM for eight days, MERTZ chatted up 510 strangers and grew to recognize repeat visitors fairly well. (Its scorecard wasn't perfect. The fourth most recognized "person" was a wall.) What was most astonishing and ultimately frustrating to Aryananda, however, was how eagerly people treated a robot that its maker describes as "dumber than a rock" as a *person*. Even Aryananda found herself fooled. "You would think that if I'm smart, since I'm working on facial recognition, whenever I look at the robot, I should face the robot the same exact way with the same exact expression," said Aryananda. "But if you notice my face [as] recorded by the robot, there were all sorts of facial expressions and all sorts of facial orientations."[13] She, too, fell for MERTZ.

Just as we expose ourselves to each other behind the veil of the virtual, so we leap toward intimacy at the thinnest signs of life, as Clifford Nass's lifework has shown. In one of his experiments, when a group of people wearing either green or blue armbands played games on either green- or blue-colored computers, those whose armband matched their monitor color won more games. Moreover, people expressed more solidarity with a color-matched *computer* after a time than with a human from a different team. One of the first and most famous early experiments in this vein was Joseph Weizenbaum's Eliza, a rudimentary computer therapy program created in the 1960s. Eliza's attraction to people surprised and disturbed its inventor. Some of his students, who knew exactly how Eliza worked, asked to be alone with the machine. Fastforward a generation, and now we have Laura, a software "relational agent" designed as a virtual personal trainer by Timothy Bickmore, a Northeastern University assistant professor of medicine who first created the character while working in Picard's lab.[14] Laura, a svelte brunette who looks like a soccer mom, is merely an animated character on a screen, and yet she unexpectedly fills a room with her empathy, suggestions, considerate listening, compliments, memory for detail, and undivided attention. She talks in a warm, gentle, although somewhat wooden voice in response to click-by-click multiple-choice answers from her human "client." It's an artificial, lopsided conversation, and yet oddly engaging. Laura opens by asking how you are, and if you answer "so-so," she looks stricken and says in a concerned voice, "I am so sorry to hear that." As I look over Bickmore's shoulder, he runs through a version of Laura that is used to persuade schizophrenics to exercise and take their medications. At one point, she suggests a walk. "Can you do that for me?" she purrs. When he clicks yes, she gushes, "That is awesome! We make a great team."

In one sixty-day experiment of twenty-one geriatric patients beginning an exercise regimen in Boston, those who talked to Laura walked twice as many steps as those who just got a pedometer and brochure encouraging exercise. One subject remarked, "I feel Laura, in her own unique way, is genuinely concerned about my welfare." Another said without irony, "Laura and I trust each other." More than one reported that Laura "liked" them.[15] "That was really bizarre," recalls Picard.[16] Bickmore, who is developing a PDA-based version of Laura dubbed the "wearable conscience," is quick to assert that people know that they're dealing with a computer. (As a jest, he programmed Laura so that if you ask her where she lives, she replies, "I just live in this little box.") And yet, many of those who interact with Laura eventually seem to feel that she's much more than a shadow puppet. "After you talk to her for so long, you get to think of Laura as a person," recounts one elderly user. "Strange as it may seem, it really works. Laura makes you feel like you've got a friend," wrote science reporter Bennett Daviss. "As I walk away, I am left with the uncanny feeling that I have just enjoyed meeting someone who lives inside a computer." Daviss admits that this is "slightly embarrassing."

In the affective and sociable computing worlds, the word *uncanny* comes up often. Explored at length by Freud in a 1919 essay, this complex notion involves something—such as a ghost—both familiar and unfamiliar, visible yet secret, which evokes a special brand of fear and uncertainty. The sensation is especially triggered by figures that inhabit the border between the lifeless and living, such as spirits, "waxwork figures, ingeniously constructed dolls and automata," wrote Freud.[17] He attributed the feeling to the resurrection of long-gone childhood beliefs in an animistic world controlled by our own mental omnipotence, where dolls could talk, wishes could kill, and things around us could fly. Like the student in E. T. A. Hoffmann's 1817 short story "The Sandman" who falls in love with his physics professor's "daughter"—in actuality, the old man's beautiful, masterwork automata—we eagerly overlook the empty gestures of the affective machine, seeing what we wish to see and hearing what we wish to hear. As Nathanael stares at Olympia's "fixed and dead" eyes, they seem to shine. When he kisses her cold lips, they seem to warm. "She utters few words, it's true," Nathanael tells a doubting friend. "But those few words are true hieroglyphs. . . . Only in Olympia's love do I find myself." When he discovers the truth about Olympia, Nathanael goes insane.[18]

Why are we such "cheap dates," in Sherry Turkle's words? The answer lies both within the depths of our biological makeup and at the surface of our social selves. Just as television ingeniously preys on a toddler's primitive need to orient to the shiny and mobile in her environment, so we have evolved to pay attention to signs of anger, sadness, joy, and a myriad number of other emotions around us for the good of our survival. "The human brain is so exquisitely attuned to emotion, so obsessed with it, and so good at detecting it, that even the slightest markers of emotion can have an enormous impact on how the brain behaves," says Nass.[19] By carefully dissecting our lexicon of socioemotional cues, particularly in the face, scientists can create machines that effectively, as Breazeal says, "push our buttons." Using a forty-four-unit coding system of facial expressions developed in the 1970s by psychologist Paul Ekman, researchers at the University of California at San Diego have spent more than a decade compiling an emotional databank of one thousand faces to help computers better read human emotion.[20] Building on that coding system, Picard's lab is developing a wearable "social-emotional prosthesis," or "ESP," to help autistics and computers read people.[21] The device takes video footage of people's faces and head movements, then spews out real-time graphs and pie charts that squiggle and shift, reporting whether a person being filmed is agreeing, disagreeing, concentrating, thinking, unsure, or interested. It does so using "mindreader" software originally created by Rana el Kaliouby, now a postdoctoral researcher with Picard, to help computers read human thought and emotion. A "self-cam"—a kind of metal necklace with an antennalike rod topped by a tiny video camera—feeds footage to the "mindreader." It's simultaneity tuned up.

After I put on the prosthesis, a readout on my emotional-social state began showing up on a laptop as I spoke with graduate student Alea Teeters, the self-cam's inventor. It was akin to watching an eerie EKG of emotion. The mindreader's track record is good: research shows that it accurately identifies whether people are unsure, agreeing, or another of the six measures of emotion at least 64 percent of the time, with up to 90 percent accuracy when reading actors. Yet at one point, the ESP paradoxically reported that I was concentrating although barely interested in Teeters as she explained the device to me. And once when I shook my head "no," my "thinking" scores plummeted. The graphics changed so swiftly that I had trouble determining if the device was gauging me correctly. Could I really have been concentrating yet bored one minute, disagreeing mindlessly the next? But that is not the point. The machine reports what I am supposedly *showing* to Teeters. And perhaps this "readout" of the social me—accurate or not—is enough. After all, we often offer each other little more than tantalizingly effective empty prompts and cues, like virtual dolls programmed to ask about another's day. The empathetic murmur from a spouse thinking about his work, the seeming warmth of the saleswoman, and the insincere compliment offered on cue, all deliver comfort to us, without authentic mutual understanding. We present to the world a tapestry of externalities that may or may not match our inner emotions, and we bumble through misreadings and misunderstandings with even those we know best. If the robot can offer us no better than this in a relationship, perhaps that's enough. "Does our knowledge that another human has far more bubbling beneath the surface than their smiles and frowns suggest allow for a more genuine interaction?" asks British social anthropologist Kathleen Richardson, who studies human-robotic bonds. "Or are a few smiles and frowns all we need?"[22]

That's all we need at times, insist Breazeal and Picard. The essential argument for affective computing is that such robots or software agents can offer an encouraging word, a comforting gesture, or a bit of advice as well or even better than humans. They may not be any better than we are at the game of social relations, but they nevertheless can be there for us when we aren't there for each other, "whether due to a dearth of others available to help, isolation, impaired socioemotional skills on one side or another, or a multitude of other reasons," write Picard and roboticist Jonathan Klein in a paper that addresses some of the philosophy and future implications of the field.[23] Klein relates an experience that inspired him to do this work. While on his honeymoon, he discovered two dead insects in an airline supper, received no apologies, and wound up with a substitute meal that gave him an allergic reaction because it was erroneously labeled dairy-free. Turning from one unsympathetic airline agent to another after the plane landed, he recalled thinking, half in jest, "I can build a *computer* that handles people's feelings better than this." While insisting that affective technologies shouldn't replace humans in society and acknowledging the "troubling" nature of offering "an idealized simulation of real empathy, real understanding and real caring," Picard and Klein, nevertheless, argue for moving forward. The field, they conclude, has only begun to "scratch the surface" of discovering how affective technologies can meet people's emotional needs, they conclude.

Are we entering a post-uncanny world? By the time we create devices that pass the Turing test and become indistinguishable from humans, we will no longer care that we have been duped, says Sherry

Turkle. A wide-eyed, petite, and fiery MIT professor, Turkle is yin to Picard and Breazeal's yang, a philosophical sparring partner in the burgeoning social debate over what it means to be human. Just back from Australia, Turkle settled into a booth across from me at a noisy diner on Manhattan's Upper West Side one spring afternoon and wearily ordered a cup of Earl Grey tea.[24] But she came to life as she began telling stories here and at a later lecture of our deepening relations to machines. She recalled visiting an exhibit on Darwin with her teenage daughter at the American Museum of Natural History and seeing two Galapagos turtles caged at the entrance of the show. Feeling sorry for the turtles and completely unmoved by the wonder of their presence, Turkle's daughter remarked that the museum could just as well have used robots. (Perhaps she was unaware that some of the first cybernetic machines were walking, dancing light-seeking tortoises built by W. Grey Walter in 1950.) Other children in line agreed, to their parents' dismay. Intrigued, Turkle returned again and again to interview visitors to the exhibit and found that for most children, "aliveness doesn't seem worth the trouble, [and] seems to have no intrinsic value." Moreover, if a realistic robotic turtle was used instead, the children didn't think people needed to be told.

Turkle wasn't surprised. For more than twenty years, she's been studying our relations with robotic machines, from needy Tamagotchi keychain pets and owlish "talking" Furby toys to new robotic companions increasingly used in nursing homes. At one of several Massachusetts nursing homes where Turkle is studying the use of sophisticated robotic seals from Japan as pets, a lonely seventy-two-year-old woman, saddened by her son's refusal to visit her, stroked the robot, Paro. "You're sad, aren't you?" murmured Ruth. "It's tough out there. Yes, it's hard." Ruth is comforted by the robot, which knows five hundred words, can make eye contact, recognize an owner's voice, and sense whether it's being touched gently or aggressively. But "what are we to make of this interaction when it happens between a person and a robot who . . . ," Turkle caught herself and stopped, "*which* has no idea what is going on?" I winced, thinking about the numerous little robotic dogs I have bought my eleven-year-old daughter, hoping to deflect her yearning for a puppy.

For the moment, Domo and RoCo are as caged as Galapagos turtles, yet they symbolize the quiet shift that Turkle notes is already playing out in daily life: as emotional machines, we are no longer alone. Our machines are built to reach for us and beg for our attention, says Turkle. "The sight of children and the elderly exchanging tenderness with robotic pets brings science fiction into everyday life and techno-philosophy down to earth," she writes.[25] If the virtual is the new "real" and robots are the new "alive," then we're already well on our way toward answering the question of whether Domo is a tool or a person. It's too late to think about scratching the surface. We are going deeper every day. And the ultimate answer will come not from the ever-expanding capabilities of our machines but from how we envision ourselves in the future, says Turkle. "What kind of people are we becoming as we develop increasingly intimate relationships with machines?" she asks.

This is the question that Czech satirist Karel Čapek sought to raise in the 1922 science fiction play that introduced the word *robot* to the world and won him international renown.[26] In *R.U.R.: Rossum's Universal Robots*, scientists eager to create the perfect worker discover how to build robots that are so like humans in behavior, intelligence, and appearance that a visitor to the robotics factory cannot tell them apart from people. The robots, however, have no soul and so are content to work as slaves, until one scientist secretly begins trying to make them human, and in doing so, unwittingly makes them discontented. The robots rebel and slaughter humanity, a run-amok climax that has been a staple of science fiction since Mary Shelley's popular 1818 novel *Frankenstein, or The Modern Prometheus*. Audiences of Čapek's time were thrilled and horrified by his "hair-raising" drama and by the idea of the mechanical man, and yet, Čapek was ultimately disappointed because he felt that people, in their fascination with the robot, had overlooked his true aim of using robots to comment on the nature of humanity. Čapek's robots were intended as a mirror of our own misguided aims, yet audiences saw in them an awe-inspiring achievement. "If we have some perfect and shining machine before our eyes, we see in it man's triumph," wrote Čapek some years later. "To realize man's defeat, we only need to set beside that splendid machine the first beggar we meet."[27] Čapek tried but failed to show us that we need not fear the robot. Rather, we should fear ourselves.

So who are we becoming as we play god to the machine and face a future of sharing our earth with mechanical creatures that we can create, love, and nurture? This question is entangled in another way with the fate of the machine. Put simply, as machines seemingly become more human, we are becoming in many ways more like our machines. Our evolutionary calling cards are shifting quietly from *Homo sapiens* to posthuman, with the tick of a pacemaker, whir of a brain implant, rush of an attention-enhancement pill, or the beep from a chip-implanted arm as we pass through security. This physical fusion with the machine empowers us and yet risks narrowing us in ways that may be hard to imagine.

Of course, we've always been at one with our hammers and hoes. As historian David Nye points out, we've used tools for four hundred thousand years; they are inseparable from human nature and are more often the product of cultural desire than raw need.[28] We push forward the boundaries of our own evolution by using objects that augment and extend our own bodies, as thinkers from Freud to McLuhan have observed. The telephone is an "artificial ear," invented by a man, Alexander Graham Bell, who was partially deaf, notes media theorist Friedrich Kittler, who also calls the telegraph an "artificial mouth."[29] Recall the close parallels in chronology and intent between telegraphy and spiritualism, both Victorian-era creations celebrated at the time as scientific ways to conquer time and distance and extend the senses. Consider the telescope, eyeglasses, wheelchair, Walkman, and iPod. As the eternal Homo faber, or man the builder, we have used all manner of tools to leverage our birth powers. Even our bodies have served as canvases for change. Electricity not long ago was seen as medicinal, magical, and an emblem of technological progress that could be experienced through the "touchstone" of the body, as historian Carolyn Marvin notes. Eighteenth-century partygoers in London and Paris liked to link hands to channel electrical impulses, believing that this encouraged rapport and communication. The controversial healer Franz Anton Mesmer, inspiration for the word *mesmerizing*, encouraged the pastime. A century later, the *Electrical Review* promoted the "electric cocktail," in which a current was used to carbonize the sugar in a drink, as an up-and-coming winter beverage of 1885.[30] In 2004, Microsoft quietly patented the human body as a "personal area network," or "computer bus," hoping to one day use the skin's conductivity to transmit data.[31] As Nye observes, our increasingly technological society has inspired us to see the world "less as a shared dwelling than as a stockpile of raw materials."[32] And now that stockpile increasingly includes our own flesh.

Dr. John Halamka cheerfully rolled up his shirtsleeve as if he was about to get a shot and revealed an arm seemingly like any other.[33] No bump or bruise, scar or sign of the cyborgian machinery beneath his skin could be seen. But when he passed a handheld scanner over his arm, he triggered a beep and a flash of numbers on the screen. Halamka, the effervescent chief information officer of Harvard Medical School and Boston's Beth Israel Deaconess Medical Center, is "chipped." In late 2004, he became the first doctor in the United States to get an RFID—radio frequency identification—chip implanted in his body. The tiny VeriChip, which is enclosed in an unbreakable glass capsule, was injected about one-quarter to one-half inch into his body by using a knitting needle–sized instrument. When scanned, an identification number is produced that can be used to access Halamka's Web-based medical records—information that is sometimes inaccessible if a patient is gravely ill, cognitively impaired, or simply unconscious. Fundamentalist Christians have decried the device as a tool of the Antichrist; Halamka regularly gets e-mails accusing him of carrying the "mark of the beast" described as a tool of the Antichrist in Revelation 13 ("And he causeth all, both small and great, rich and poor, free and bond, to receive a mark in their right hand, or in their foreheads").[34] Privacy advocates are wary too, noting that the so-called prosthetic biometric is not encrypted and so easily could be used for anonymous tracking and identity theft. Anyone with a hand-held scanner, for example, could steal a stranger's data and make a duplicate chip—a new twist on the data-falsifying crime oddly called "spoofing."[35] Halamka, who has no ties with the manufacturer, writes and speaks openly about these risks, matter-of-factly admitting that the grain of rice-sized morsel of silicon that he carries within him has destroyed his privacy. (The chip, however, can be removed with minor surgery and the help of an X-ray to locate it.) "I've learned that I've lost my anonymity completely," says Halamka. A slim, dark-haired speed-talker, he bounced around his Boston office dressed all in black, showing me a sample chip, a bit of audio on his laptop, a souvenir from Japan. Almost gleefully, he recounted numerous episodes of setting off alarm bells at various stores.

As an emergency room physician and technologist who leads a national nonprofit body responsible for overseeing US medical data standards, Halamka said that he felt compelled to get chipped in order to fully investigate the technology. Was it painful? Would he forever feel something in his arm? How much of a privacy compromise was he making? He had to do it to know. In a sense, he turned himself into a piece of the information that is his calling. A man who's responsible for data involving 3 million patients, 14,000 employees, and 3,000 doctors and standards for the nation can now be scanned and read. Persuasively, he argues that he didn't do it for the hype, and yet you can't help but feel that Halamka, who's also signed on to be one of the first to post his complete genome online, welcomes the attention provoked by his new cyborg self. Halamka likes to live on the edge, certain that he shapes his destiny. He dropped seventy pounds after his doctor told him he was nearing obesity and now is a caffeine-shunning, vegan mountain climber, proud of his ability to steamroll through six hundred e-mails daily and to scale high peaks with a BlackBerry at his belt. He's at the forefront of something we can't all fully understand but find intoxicating—the promise of going beyond augmentation to a redesign of the human through a melding with the machine. Is the chip "dehumanizing" as some of Halamka's colleagues assert?[36] That might be the point in a posthuman world. "We will have the power to manipulate our own bodies in the way we currently manipulate the design of machines," predicts AI pioneer Rodney Brooks.[37] Frank Moss, head of the MIT Media Lab, is more blunt. "We're hacking the human!" he exulted at a packed event to unveil a new mission for the lab—creating "Human 2.0." We're shedding our old selves like snake-skins in the grass.

Consider all things "prosthetic." The word, which derives from the Greek for *addition*, first entered the English language in 1553 to desig-nate the addition of a letter or syllable at the beginning of a word. Only in the early 1700s did the term begin to be used to indicate a surgical remedy for a deficiency, such as an artificial tooth or limb.[38] Prostheses were sometimes elegant yet nevertheless dead wood appendages, second-tier fill-ins for the glory of our biological selves. But now the prosthesis has come to life, both literally and, in critic Vivian Sobchak's words, as a "sexy new metaphor" for a "vague and shifting constella-tion of relationships among bodies, technologies and subjectivities."[39] Advanced prosthetic limbs, such as the German-built C-Leg, are now sophisticated wearable robotics that take charge of human movement. A tapestry of sensors that take fifty measurements a second feeds into a microprocessor that controls the C-Leg's hydraulics system, just as a brain controls our muscles. This technology alone helps transform shame into glamour, turning the wearer into a cyborg and capturing our imaginations with visions of bionic, extended, and ever-mutable selves. "The posthuman view thinks of the body as the original prosthesis we all learn to manipulate, so that extending or replacing the body with other prostheses becomes a continuation of a process that began before we were born," writes literary scholar N. Katherine Hayles in *How We Became Posthuman*.[40] This is why the charismatic blonde supermodel and para-athlete Aimee Mullins, a bilateral amputee below the knees, is the poster girl of prosthetic-chic. She has broken national and world records in track as a disabled athlete, graduated from Georgetown's School of Foreign Service, and she switches prostheses the way some change jewels. (She has different sets of legs depending on whether she wants to wear flats, or inch-high, two-inch, or four-inch heels.) Mullins shows us "an unstoppable 'difference' that is not about negation but about the alterity of 'becoming,'" writes Sobchak, who wears a prosthetic leg due to cancer. Those of us who inhabit only flesh can wear glass slippers. Mullins can slip on glass feet. The body isn't meat. It's a sculpture we can mold.

And no prosthetic realm is more alluring than the frontier of the mind. "If your brain can do it, we can tap into it," says John Donoghue, a professor of neuroscience at Brown University and leader in the burgeoning field of "neuro-prosthetics."[41] Donoghue is talking about the brain implant he developed that allows severely paralyzed people to control computers, televisions, or robots by using only their thoughts. By 2007, four people had received the BrainGate system, which is made by a Massachusetts company that Donoghue cofounded to bring fifteen years of laboratory work with monkeys into the human realm. BrainGate, which is essentially an artificial nervous system, involves inserting a baby aspirin–sized sensor with a hundred microelectrodes into an area of the motor cortex. Once connected by cable to a computer, the sensor eavesdrops on milliseconds-long neuronal electrical signals called action potentials that Donoghue calls the "brain's language of information." Merely by thinking about moving a limb, a paralyzed person with an implant can send signals into a computer for translation into actual movement of a cursor, remote control, or prosthesis. "We listen in on the conversation of the brain," says Donoghue. In the not-distant future, he predicts that epileptic seizures or bouts of depression will be stymied before someone is even aware of a nascent episode, or the system will be linked with muscles or a prosthesis so that people could walk or gesture just by thinking about it. Yet future applications go beyond the purely medicinal. "Supposing you had a person quite elderly who didn't have much longer to live, and couldn't get out of bed," muses Donoghue, a kindly, debonair man with a Santalike shock of white hair and beard. "What if you put an implant in their head, and allowed them to manipulate the world around them, giving them an altered reality? What if you could experience that at the end of your life? That will happen." Donoghue is quick to say he isn't advocating the technology; but he predicts that it's coming.

Perhaps a pill would do the trick as well. Another emerging realm of neural prosthetics is pharmacological, mainly involving drugs that augment or shape our powers of attention and memory—and potentially our identity. For if any two cognitive capabilities lie at the core of our identities, they are attention and memory, whose mysteries have only recently begun to be unlocked. Memory is a "rearview mirror on life," according to memory researcher James L. McGaugh. "We have this magical machine in our heads that captures the experiences we have and retains them."[42] Attention, in turn, is a window onto the world, a key to shaping our lives and our environment. No wonder that the incorrigible Homo faber has begun an explosion of experimentation, scientific and individual, into manipulating these cognitive powers. There is Adderall and "Vitamin R," or Ritalin, used to boost focus by college students and teens with both self-described and diagnosed attentional deficiencies. Modafinil, a drug approved in 1998 for narcolepsy, is now used by the military to enhance wakefulness without the jitters produced by traditional stimulants. The allure is obvious in a sleep-deprived culture. Buoyed by off-label use, sales exceed $575 million yearly. "If I take a dose just before I go to bed, I can wake up after four or five hours and feel refreshed," said a thirty-something software developer who has been using modafinil after getting diagnosed online with narcolepsy. "I find I can be very productive at work. I'm more organized and more motivated," he told *New Scientist* magazine.[43]

Modafinil is a star of the new realm that neurologist Anjan Chatterjee calls "cosmetic neurology," but dozens of such pills are waiting in the wings, all of which manipulate brain chemistry. If Donoghue's work is a kind of listening in on and translating talk between the brain and body, then these drugs hold the potential to change cognitive chat itself. Whenever we think, feel, move or remember, flows of ions and neurotransmitter molecules are released that carry information between neurons. This is essentially the "conversation" that underlies learning and thought. One class of new memory enhancers called ampakines boosts the frequency and strength of neuronal signaling by amplifying the key neurotransmitter glutamate. Other discoveries by Nobel Prize winner Eric Kandel and scientist Tim Tully have led to efforts to boost CREB, a protein involved in switching on and off genes involved in forming memories. And in a first clinical effort to create a kind of forgetfulness, Harvard psychiatrist Roger Pitman has built upon McGaugh's research to study how the hormonal beta-blocker propranolol can diminish traumatic memories of crime and accident victims, 15 percent of whom on average go on to develop post-traumatic stress disorder.[44] Emotion etches experiences into your memory because adrenaline and other stress hormones that accompany a moment of joy, fear, or other arousal amplify the chemical process of memory consolidation. (In medieval times, a child chosen as the official observer of an important event was afterward tossed into a river, a terrifying excursion that made him long remember the events of the day, McGaugh notes.)[45] Pitman isn't quite offering what fictional characters from the ex-lovers in *Eternal Sunshine of the Spotless Mind* to *Macbeth* have dreamed of: to "pluck from the memory a rooted sorrow, raze out the written troubles of the brain."[46] Instead, he's essentially *stunting* the growth of a memory just planted. Yet a day-after pill for our bad times is, nevertheless, heady stuff. "Propranolol might be the most philosophically vexing pharmaceutical since Prozac: it openly questions the significance of reality," writes pop culture commentator Chuck Klosterman. "This seems wonderful and terrifying at the same time."[47]

Indeed, redesigning the human inspires panicky visions of ever-content, soma-chomping citizens on one hand, and heady "transhumanist" promises of living forever, free from sadness, disease, and ignorance on the other. Enhancement will liberate us from "the sick psycho-chemical ghetto bequeathed by our genetic past" and offer the eradication of all suffering, proclaims philosopher David Pearce.[48] In contrast, conservative biomedical philosopher Leon Kass argues that enhancements, especially pharmacological, are wrong because they are unnatural and unfair. "Nothing hurts if nothing matters," writes Kass, comparing a future of drug-induced contentment to the stupor enjoyed by denizens of *A Brave New World*.[49] Between these two extremes lie a host of unavoidable and often age-old questions involving social inequality, ethical fairness, and medical safety. For example, only the rich might get enhanced, so that, as Katherine Hayles surmises, "having an unmodified body will be like having a working-class accent; it will mark you as cannon fodder for the system."[50] Some argue that short-cut improvements constitute cheating and that enhancements may prove a Faustian medical bargain long-term. "We understand very little about the design constraints that were being satisfied in the process of creating a human brain," argues neuroscientist Martha Farah, who studies the effects of modafinil in non-narcoleptics.[51] Still, much of the debate about our new fusion with the machine misses an important point: we tend to forget that our chips and implants and pills effect powerful changes within the broad *context* of our humanity and our environment. Again, who we are and who we want to be matters more than what our ever-improving machinery can or cannot do. This is the gentle warning that neurologist Oliver Sacks offers when he tells the story of "Virgil," a blind man pressured by his fiancée to have surgery to restore his sight. Afterward, his eyes work, yet he is unable to see anything more than disjointed colors and shapes. In part, this is because seeing is a complex, learned skill. But the restoration of sight, as well, shattered a world that Virgil had navigated with surety and understanding. The blind often have extraordinary powers of "vision" from using a kind of sonar orientation. At the same time, the sighted are often blind to much of life. Summing up the tale of Virgil, Oliver Sacks quotes philosopher Martin Buber: "We must humanize technology before it dehumanizes us."[52]

In his poignant memoir about receiving a cochlear implant, *Rebuilt: How Becoming a Cyborg Made Me Human*, Michael Chorost tells the story of his own slow realization of the limits of technology to transform humanity. Born almost completely deaf, he spent a lonely childhood and early adulthood feeling on the fringes of society. He was a geek who became disenchanted with the relentlessly abstract logic of computers, exemplified by online dating systems built to make nonoptimal types like the 5'4" Chorost virtually invisible. The computer "offered no social context in which I could be judged as a human being," he writes.[53] Nor could technological devices make him *human*, as he began to discover one day during a group therapy session. After a self-concocted amplification system for his hearing aids broke, he was startled to realize that he could still hear the other participants talking. Previously, he'd been too focused on his own troubles to attend to others, through his senses or his heart. "It was then that my technology fetish had begun to fade. The gratification of upgrading my technology diminished. Upgrading *myself* was hard, slow, subtle work, but it had the enduring satisfaction of true craft."[54] Chorost ultimately exults in the mutability of the cyborg self, yet rejects the idea that the technologies fusing man and machine will automatically expand human powers. Furthermore, he questions whether or not we are losing our capabilities for growth and self-mastery as we look to our machines. "The one hundred and forty thousand transistors in my skull give me sound, but they cannot make me *listen*," he concludes. "It's only when I listen that my cyborg technologies make me a better human being."[55] Having a prodigious memory cannot help us separate trivia from wisdom. Having data implanted in our arm can't tell the world who we really are. As technology critic Steve Talbott wisely reminds us: "Self-forgetfulness is the reigning temptation of the technological era."[56] The body isn't meat, it's the reservoir of our humanity. And there will never be a prosthesis for the human spirit.

"As the great stone circle at Stonehenge reminds us, [tools and machines] are part of systems of meaning, and they express larger sequences of actions and ideas," says David Nye. "Ultimately, the meaning of a tool is inseparable from the stories that surround it."[57]

Moreover, storytelling and tool use are similar processes, explains Nye. Each requires imagining altered circumstances, controlling an outcome, and the passage of time. "It seems likely that storytelling and toolmaking evolved symbiotically, analogous to the way that oral performances are inseparable from gestures and mimicry." So what stories are we weaving as we look to the machine to comfort and transform us—indeed, to be a part of us? Within this messy convergence, we are on the brink of redefining humanity, but in ways that ultimately may impoverish us. In a distracted time, our virtual, split-screen, and nomadic lives nurture diffusion, fragmentation, and detachment. We begin to forget how to pay attention to one another deeply and begin to attend more to fallacy and artifice. Trust, depth of thought, and finally a certain spirit of humanity begin to be lost. Such changes are harbingers of a wildly inventive, marvelously technological dark age.

Consider prosthetic chic and the technologies that are already allowing us to play at the edges of a cyborgian identity. Much current experimentation operates within the unarguably positive realm of helping the disabled or the sick. An implant to help a paralyzed man connect virtually to others is a remarkable step forward. A pill to reawaken the memory of someone with Alzheimer's could save many from the disease's multitudinous cruelties. Yet once we begin to consider cases that are less about medicine and more about augmentation, the issues grow murky. Our relentless ancient drive to improve our technologies, as Nye points out, is often not about *need*, but about social evolution, that is, what we *value* and how we define need. In the not-distant future, we will face an array of enticing appendages—from prostheses-for-all to focus-enhancing pills—that can bolster us with push-button speed. And we likely will face these alluring choices in an era of virtuality, time droughts, split-screen living, and diffusion. If we opt for the ever-increasing upgrade, will we continue to value the messy limitations of the body? Will we dread the hard work of mastering ourselves and choose to simply pattern ourselves fully after the machine? When we hear better, will we listen? The film critic Vivian Sobchak, who lost a leg to cancer, expresses admiration for Aimee Mullins's ability to spin a public body image based not on deficiency but on fashionista mutability. But Sobchak says she'd rather treat her prosthesis as simply a means to get around, not the source of her identity. "All I want is a leg to stand on, a limb I can go out on—so I can get about my world with a minimum of prosthetic thought," she writes.[58]

Man into machine, and machine into man. Consider, as well, the implications of giving the old, the young, and the sick the companionship of a sociable robot or a virtual trainer. Such devices can and do provide guidance and comfort, as the elderly patients' praise for Laura can attest. However, we must remember that affective machines elicit feeling from us yet have no real emotion. "What is the value of interactions that contain no understanding of us and that contribute nothing to a shared store of human meaning?" asks Turkle, adding that robotic companions may eventually desensitize us to the kind of attention that only people can give each other. Socially, humans often deceive one another—offering smiles that hide anger, comforting words that mask an absence of care—yet inherent in our interactions is the *potential* for mutual understanding. There is no such hope if you love a robot. Moreover, if we turn to machines when we can't be there for one another, we risk losing the will to *keep trying to fully connect.* "We had better be careful just what we will build, because we might end up liking them, and then we will be morally responsible for their well-being. Sort of like children," warns AI pioneer Rodney Brooks.[59] Spend a bit of time with Laura, and you begin to realize how right Brooks is. In an era when we show signs of becoming content with thinner relationships, it may become easier for us to care morally for the machines that we in turn program to care for our most vulnerable members of society. And what a distraction that would be from our responsibility to one another.

Before one of my treks to MIT, I made a very different pilgrimage up New York's Hudson River to a small town at the foot of the Catskill Mountains to visit a man who's been wrestling quietly with such questions for decades.[60] Steve Talbott is an Oregon-born, former organic farmer, who has developed a growing following for his critical writings on technology—a body of work that the *New York Times* calls a "largely undiscovered national treasure." Talbott carries on in the tradition of some of the great critics of technology: Jacques Ellul, Lewis Mumford, and Neil Postman. Yet he's distinct among such thinkers in that he worked for years as an impassioned computer programmer and technology writer. Now he's a second-career contemplative, living a quiet, country life and driven by a new cause: to make the connections we can't see, sound the warnings we are too deaf to hear. In his long-running online newsletter "Netfuture," his books, and talks, Talbott argues that our dependence upon and identification with our machinery is leading us to sleepwalk into a dehumanizing future. Yet Talbott is no born-again Luddite, smashing his laptop in the barn, but rather a firm believer in living with our machinery without losing ourselves to its charms.

Talbott picked me up at a local train station, and we drove in his ten-year-old stick-shift Volkswagen Golf back to the Nature Institute, a nonprofit where he now works. Dressed in jeans and a dark green mackintosh, he had a careworn look. A bout of illness has left him gaunt and, by his own good-humored admission, looking older than his sixty-something years. But he has a firm, deep voice that matches his convictions and a startling way of pointing out the hidden "systems of meaning"—the stories—that underlie our technological world. Computers are wondrous tools that are and should be a part of us, says Talbott. Yet we quickly forget that such miracle machines spring largely from the human potential for analysis, mechanization, and cold logic, in other words, from just one lopsided slice of humanity. We had to "enter into our own potentials for programmed, automatic thought and action before we could build automatons of silicon, plastic and metal," writes Talbott in *Devices of the Soul: Battling for Our Selves in an Age of Machines*.[61] And when we become smitten with this narrow type of thinking, we begin to lose sight of our own far-greater inner capabilities and put our machines on a par with ourselves. We cease to see our machines as separate from us and pattern ourselves after them. "That is why we so readily give our assent to the absurd proposition that a computer can add two plus two, despite the fact that it can do nothing of the sort—not if we have in mind anything remotely resembling what *we* do when we add numbers," Talbott reminds us. "In the computer's case, the mechanics of addition involve no motivation, no consciousness of the task, no mobilization of the will, no metabolic activity, no imagination."[62] The fact that this simple truth may be so startling is a measure of how deeply we believe in the notion that our power and potential lies with the machine.

Our liberation must come from overcoming this fallacy, said Talbott, pecking at a salad that we picked up at the local farm store, his silhouette framed by a view of rolling foothills. To explain, Talbott often returns to the root of our word *technology*, the Greek *techne*, meaning craft, skill, cunning, art, or device.[63] This double entendre, *techne* as both a skill and a device, is echoed in the Greek word *mechane*, which is the source of our words *machine*: a physical construction, and *machination*: a contrivance. Talbott is too cunning a student of *techne*, in its ancient and modern senses, to simply take the cheap shot of pointing out this intriguing etymological association between technology and deceit. Rather, he revisits the story of Odysseus, tenacious voyager and "trickster par excellence," to explore the difficult balance between the use of clever devices and the resourcefulness needed to use them wisely. Before Odysseus passes the entrancing song of the Sirens, for example, he orders his sailors to plug their ears with wax and lash him to the mast so that he could listen to their false promises of endless wisdom but not act upon them. Odysseus, in effect, counters their devices with one of his own, through his clever self-possession. Today, however, we have lost the necessary understanding that an inner resourcefulness must always go hand in hand with using our devices, or else we risk ceding control of our lives to our tools. Only then can we differentiate ourselves from our machines and realize how incomplete they are in their beautiful, cold, analytical narrowness. Only then will we realize how melding ourselves with our machines is ultimately a gesture of "hope, confusion and pain," a misplaced dream and simultaneously a defeat.[64] Talbott's words remind me of Tim Bickmore's occasional discomfort with the future impact of his own creations. "I try to always justify it by saying that I'm making people healthier," said Bickmore, who is also developing a disembodied robotic glove to hold hands with hospital patients. "But still, there may be even longer-term issues than one individual's life time to worry about." As we talked, Laura kept eerily intruding. "Are you there?" she asked plaintively. "Have you seen any good movies lately?" Finally, ignored, she shut off with a cheerful "Good-bye." Čapek was right: we have only ourselves to fear.

So is Domo a tool or a person? A tool, if Aaron Edsinger has his way. He has started a robotics company, but he won't be making any robotic pets. I asked if he would give such a creature to his mother, and he unhesitatingly said no. It's all very interesting, the power that you wield when you push those buttons, said Edsinger in his unhurried beatnik style. "But it's not something that I'm completely comfortable with," he said. He envisions future Domos as appliances—handing you the salt, clearing the table, answering the door. As we talked, he demonstrated the robot's crude domesticity, telling it to grasp a box of crackers and fill a cardboard tray with empty cups and packages. A bit envious of Edsinger's fun, I asked to touch the robot, then pushed and pulled on its arm, vainly hoping to elicit an "ouch." But once, as I reached out to touch it, Domo firmly took my hand. "Domo, give it," said Edsinger, ordering the robot to let go. Instead, Domo pulled my hand steadily toward its body. "Domo, stop," said Edsinger. It parroted, "Domo, stop" but kept tugging. Domo wouldn't let go. Finally, I extricated my hand gently from its grasp, bid its maker good-bye, and left, the sensation of Domo's insistent clasp lingering on my skin as I headed home.

Part III

DARK TIMES...

Or Renaissance of Attention?

Chapter Eight

McTHINKING AND THE FUTURE OF THE PAST

The place where past meets future—the present—is a point of eternal struggle for man, as the philosopher Hannah Arendt depicts in *Thinking*, the first volume in her two-part series *Life of the Mind*. "The present, in ordinary life the most futile and slippery of the tenses—when I say 'now' and point to it, it is already gone—is no more than the clash of a past, which is no more, with a future, which is approaching and not there yet," she observes, drawing from allegories on time by Kafka and Nietzsche.[1] Man pushes forward into the future to escape the weight of the past and is driven by fear of impending uncertainty back to emulating the comforting reality of what's been. It is thinking, however, that offers a quiet eye in the storm of time, stresses Arendt. In thought, man can find meaning and judge the affairs of humanity, "never at a final solution to their riddles but ready with ever-new answers to the question of what it may be all about." Echoing Arendt, renowned futurist Paul Saffo writes in the *Harvard Business Review* that forecasting the future is really just the delicate delineation of life's uncertainties, helped by consistently glancing backward twice as far as you look ahead.[2]

Are we heading into a dark age? To ask this question is first to wonder whether we at present have much of a collective appetite for wrestling meaningfully with uncertainties, and whether we have the will to carve out havens of deep thinking amid the tempests of time. To deepen the riddle, note that there are likely as many definitions of a dark age as there are lost civilizations buried beneath the earth's shifting sod and sand. To the late urban studies guru Jane Jacobs, a dark age is a "cultural collapse" that leads to an "abyss of forgetfulness."[3] Anthropologist Joseph Tainter treats a dark age as simply a decline in literacy, a kind of minor player in the larger economic and political drama of collapse.[4] Perhaps no two scholars holding a mirror to the past will see quite the same reflection.

All agree, however, that amid the darkness, pockets of light still shine. "A dark age is not merely a collection of subtractions," asserts Jacobs.[5] And surely, we live in a time of wonders: we can be anywhere on earth within hours, peek into the living architecture of the brain and body, access avalanches of information with a simple point and click. In a few generations, radically new experiences of time and space have become banal. During a two-week stay in Norway, my daughter, then aged thirteen, called home one day by cell phone from a mountaintop. My husband thought at first she'd been hurt, but she simply wanted him to resolve a midhike teen debate about some Beatles' lyrics. Our instant-access abundance on many scores is alluring and comforting. Yet as the twentieth-century British political thinker John Strachey once noted, "the sunset colors of a civilization are the most lovely."[6] No one is going to write *dark age* on our walls, at least not in plain, clear script. We'll have to see for ourselves. In her essay collection *Men in Dark Times*, Arendt asserts that the bright spots in the darkest of times "may well come less from theories and concepts than from the uncertain, flickering, and often weak light that some men and women, in their lives and works, will kindle." She continues: "Eyes so used to darkness as ours will hardly be able to tell whether their light was the light of a candle or that of a blazing sun," but that evaluation can be "safely left to posterity."[7]

Mesmerized by the flickering charms and lightning-fast shifts in our own time, perhaps we can't tell at first glance whether what's creeping around us are rippling shadows or a fearful twilight. But while the final assessment of individual goodness perhaps can be left to posterity, we can't wait that long to ascertain our fate. To fail to do so is to surrender to the embrace of distraction. If we want to shape our own future, we must consider how we want to live and how we want to define progress, and as we do so, prepare to welcome to our ranks the thinking person's most prickly yet necessary companion—doubt. Observes Daniel Boorstin, "The greatest menace to progress is not ignorance, but the illusion of knowledge."[8]

Do we yearn for such voracious virtual connectivity that others become optional and conversation fades into a lost art? For efficiency's sake, do we split focus so finely that we thrust ourselves in a culture of lost threads? Untethered, have we detached from not only the soil but the sensual richness of our physical selves? Smitten with the virtual, split-split, and nomadic, we are corroding the three pillars of our attention: focus (orienting), judgment (executive function), and awareness (alerting). The costs are steep: we begin to lose trust, depth, and connection in our relations and our thought. Without a flourishing array of attentional skills, our world flattens and thins. And most alarmingly, we begin to lose our ability to collectively face the challenges of our time. Can a society without deep focus preserve and learn from its past? Does a culture of distraction evolve to meet the needs of its future? These surely are litmus tests of a new dark age and challenges we look perilously at risk of failing.

> The power of the memory is prodigious, my God. It is a vast, immeasurable sanctuary. Who can plumb its depths?
> —St. Augustine, AD 354–430[9]

The prolific St. Augustine had to grope for words to describe the magic of memory. Along with a sanctuary, he likened it to "a great field or a spacious palace," and a "storehouse." Memory is a place, yet far more than a place, incomprehensible in its size and workings. In darkness and in silence, Augustine marveled that nevertheless he could picture colors or hear music. "I can sing as much as I want, even though my tongue does not move and my throat utters no sound," he wrote. "All this goes on inside me, in the vast cloisters of my memory. In it are the sky, the earth and the sea, ready at my summons . . . In it, I meet myself as well."[10] This was why the Muses—Calliope, Euterpe, Erato, Melpomene, Thalia, Terpsichore, Polyhymnia, Clio, and Urania—were the offspring of Mnemosyne, offering to humankind the threads with which to weave all creations: stories, poetry, art. "It is through the Daughters of Memory that Homer remembers how to make a sacrifice, how to launch a ship—those word patterns that repeat over and over again through the *Iliad* and the *Odyssey*," writes Clara Claiborne Park. "The Muses were singing in his ear. The Muses are *how* he remembers."[11] Nowadays, however, we rely on store-houses of memory that are attached to us, not within us; the Black-Berry, cell phone, laptop, and even the book bounce jauntily upon our hips, ready at our summons. The virtual world, in turn, offers a kind of prefab alternative version of the vicarious imagery that we once sought within. Like Augustine, we can sit in a room and see the whole world go by, but we need no longer journey within ourselves to find it. "All of us resist the work of memory," laments Park.

And yet at the same time, we fear the failings of the "magic show" of memory.[12] The vaguely sinister blotting out of recollection by amnesia, the terrifying abyss of forgetting produced by Alzheimer's, the gaps in experience sown by the onset of aging hold a terrible fascination for a society drowning in information and desiring all knowledge at an instant. Memories can "crash," computerlike, or be malevolently wiped away, a plot line recurrent in modern science fiction from *Brave New World* to the film *Dark City*. In Huxley's book, the Controller waves his hand, "and it was as though, with an invisible feather whisk, he had brushed away a little dust, and the dust was Harappa, was Ur of the Chaldees; some spider-webs, and they were Thebes and Babylon and Cnossos and Mycenae. Whisk. Whisk—and where was Odysseus, where was Job, where were Jupiter and Gotama and Jesus? . . . 'History,' he repeated slowly, 'is bunk.'"[13] Perhaps our uneasiness stems from unconsciously sensing what Augustine knew well: that human memory isn't just an old person's comfort, or a cognitive storehouse of dusty snapshots. Memory is a "living thing," observes Eudora Welty, the "greatest confluence of all."[14] Memory in all its glorious layering evolved in humans not just as a "rearview mirror" on life but as a kind of predictor of the future, says neuroscientist James McGaugh, since whoever could remember the placement of a bear's cave or a fishing hole survived to shape the future. Ghostly, mysterious, and inextricably linked to attention, memory is a key player in the eternal battleground between past and future. To understand how, consider why we are built with a need not just to remember, but to forget. Consider the case of A. J.

Since she was a child, A. J. has been able to recall nearly every detail of her life. "I can take a date between 1974 and today, and tell you what day it falls on, what I was doing on that day, and if anything of great importance occurred on that day, . . . I can describe that to you as well," wrote the then thirty-four-year-old A. J. in 2000 to McGaugh, pleading for help.[15] Intrigued, McGaugh and colleagues spent six years studying A. J. before pronouncing her the first known case of "hyperthymestic syndrome"—a term the scientists coined from the Greek for "more than normal" and "remembering." McGaugh explains, "What makes this young woman so remarkable is that she uses no mnemonic devices to help her remember things. Her recall is instant and deeply personal."[16] Before A. J., cases of superior memory had been recorded in two types of people: normal people who excel at using mnemonic devices and low-functioning savants who have pockets of often irrelevant recall, such as calculating the day of the week for any date going back sometimes thousands of years.[17] A. J. has graduated from college, worked, and married, yet even the dry scientific accounts of her prowess reveal a poignancy to her life. She aced a neuropsychological memory test given by McGaugh but has difficulty with some executive tasks such as categorizing information, one likely reason why she didn't excel in school. An anxious soul who dislikes change and obsessively keeps diaries in tiny script in date-books, she is proud of her memory yet calls it a "burden." She is chained to her past and to a memory that she describes as "a running movie that never stops." Reading accounts of A. J., it's hard not to think back to a Jorge Luis Borges short story, "Funes, the Memorious," the tale of a young man so chained to the minutia of the past that he "was not very capable of thought." Ireneo Funes not only remembered every leaf on every tree he'd ever seen but recalled each sighting distinctly, as yet another in his churning sea of memories. "To think is to forget a difference, to generalize, to abstract," writes Borges. "In the overly replete world of Funes, there were nothing but details, almost contiguous details."[18] A. J., pronounced one of her researchers, is "both a warden and a prisoner of her memories." She is Funes come to life.

The imperfections of memory that we often rue preserve us from A. J.'s fate. Forgetting—near-instantly or over long periods—is part of the brain's constant attempts to filter and comprehend the environment. And such sifting begins at the most basic cellular levels of the brain, back down in the world of the lowly neurons, whose crackling connectivity creates the music of thought and perception. All neurons have a single long signaling axon and many branchlike receptor dendrites that in a sense speak to other neurons in highly specialized circuits across the brain. As Nobel Prize-winning researcher Eric Kandel explains, an axon is akin to a pair of lips whispering close to an ear or dendrite, passing information along via both flows of ions and chemical messenger neurotransmitters. These whispered "messages" travel across the *synapse* or gap between neurons, prompting changes in the workings or even the structure of their neighbors.[19] Memories are thought to be stored within these intricately patterned synaptic connections, across the entire surface or cortex of the brain. The brain has dozens and possibly hundreds of differing messenger neurotransmitters, but one—glutamate—is a key to the chemistry of memory making in the hippocampus and cortex. Most neurotransmitters work bit by bit, firing a receptor neuron incrementally. But in the realm of glutamatergic synapses, only a large buildup of the chemical will "turn on" another neuron, causing what Robert Sapolsky describes as a "wave of excitation." A synapse that has experienced this excitation will be sensitized in the future, firing up in reaction to lower doses of glutamate. This is the root of learning. "That synapse just learned something; that is, it was 'potentiated,' or strengthened," explains Sapolsky.[20]

The image is striking. Neurons involved in the making of memory seem to need a fair bit of prodding to get going. They operate on a threshold basis, "presumably to ensure that only important, life-serving experiences are learned," notes Kandel, whose research helped uncover the molecular roots of learning.[21] After all, we can't retain everything. But when sufficiently aroused, neurons crackle with glorious ferocity, lighting up our brains with sensations from epiphanies to nostalgia. The longer and stronger the synaptic connections, the longer-lasting and stronger the memory, so much so that amid permanently robust synapses, neurons grow new communication points, that is, terminals of the axon, or receptors of the dendrite. As we saw from Bruce McCandliss's research on reading, our brains are literally changed by learning and remembering, a point Kandel makes elegantly in his memoir, *In Search of Memory*. "If you remember anything of this book," he observes, "it will be because your brain is slightly different after you have finished reading it."[22]

What drives this elegant biological dance? In the amorphous, evolving cloisters of our memory, attention is the chief curator. Certainly, a deeply shocking or highly emotional event, such as a car crash, can excite a strong flow of the hormones and neurotransmitters that etch much of an experience into our brains, flashbulb-style. But most of the time, as we saw from exploring a culture of "lost threads," memory making of any kind grows relatively slowly from repetition and practice, bolstered by a chemistry of emotion but shepherded along via attention. Take the case of a simple telephone number, cradled in our short-term, working memory as we search the house for the wandering phone. Used just once, the number likely will be soon forgotten. But during those few moments when we are trying to keep the number in mind, the executive attention network is hard at work. The central executive component of working memory, which is driven by executive attention, coordinates and directs two kinds of cognitive cupboards designed to hold the ephemera of the mind. In the parietal lobe's "phonological loop," the number is kept and rehearsed. A separate "scratchpad" temporarily handles details of the visual-spatial world, such as the piles of laundry you must step over to search for the phone. Decide to try to remember the phone number permanently, and attention is equally crucial. This small feat involves explicit memory, one of two main realms of remembering. An experienced driver or tennis player relies on implicit, automatic, and largely unconscious recall of skills and habits tucked away in the cerebellum and striatum, two regions involved in sensory perception and motor skills. These deeply held memories probably help fuel intuition, according to cognitive neuroscientist Michael Posner.[23] In contrast, explicit memory involves conscious recall of people, places, and facts stored first in the prefrontal cortex, then converted by the hippocampus into memories etched into the circuitry of the cortex, or brain's surface. Almost any fairly difficult task draws on both forms of memory, and yet to make sense of the world, we could never rely only on implicit memory. The elaborate workings of explicit memory, which are created and summoned through conscious attention, are the real magic show of memory. Such memory making is all about the rich, carefully cultivated connections that thinkers since Aristotle have understood as the roots of knowledge. Attention, you might say, is the nectar of Memory and her children, the Muses.

Today, it is not memory's vastness but its selectivity that we should admire, as literary critic Alberto Manguel points out in his tribute to the icon of cultural memory—the library. "Our mistake, perhaps, has been to look upon a library as an all-encompassing but neutral space," writes Manguel in *The Library at Night*. "Every library is by definition the result of choice and necessarily limited in its scope. And every choice excludes another, the choice not made."[24] In the realm of nature, we are built to eternally select and reject the present as it melts into the past. But are we as a society willing to do so?

Amid vast, swelling reservoirs of information, our memory keepers are tormented by visions of data droughts. A first scenario of doom involves waves of disappearing data. Our collective memory is not stolen. It evaporates—poof! Making data is child's play, but keeping it, alas, is like trying to preserve a sand castle from the tide. The headlines are dire: "File Not Found," "Digital Ice Age," "It's the End of Your Data as You Know It."[25] Laundry lists of losses expand: Web sites from 1990s presidential campaigns, raw data from early satellite probes of Mars, 80 percent of movies produced befoe 1930.[26] With most of the world's information now digital-born, James H. Billington, the Librarian of Congress, is candid about his fears that we will face losses on a scale unmatched since the destruction of the library of Alexandria, known as the Temple of the Muses. "One ghostly image haunts all of us charged with preserving the creative heritage of humanity: the specter of the great, lost library of Alexandria [around the third century]," said Billington in a 1993 speech. "There was no back-up!" he said in an interview in 2006. "And people took it for granted, until it was gone."[27]

The specter of big, black holes swallowing our cultural memory is real. Whisk. Whisk. The dark joke among librarians and archivists, true for centuries but achingly apparent only now, is that the world's best-preserved data are carved in stone (or, some might argue, DNA). Babylonian cuneiform tables beat all when it comes to fixity. The forty-five-thousand-year-old sphinx might be worn, but hey, it's still there. Unlike the wisdom set in little *dos-à-dos* books or paintings or pyramids, digital data is kept on incredibly precise and fragile media—the archival equivalent of quicksand. A report by the country's official authority on digital longevity, the National Institute for Standards and Technology, notes that CDs and DVDs might last anywhere from twenty to two hundred years, assuming they are stored properly.[28] Color snapshots become unstable in three or four decades. Digital storage tapes perhaps will keep a decade. More alarmingly, such media often do not crumble slowly across centuries, showing visible signs of decay. They easily crash, quickly and unexpectedly, because small glitches in their internal or external structure—magnetic degradation on a floppy disk, a scratch on a shiny CD—can render a copy unreadable. "We have gleaned information from letters and photographs discolored by exposure to decades of sunlight, from hieroglyphs worn away by centuries of wind-blown sand, and from papyri partially eaten by ancient insects," writes archivist Daniel Cohen. "By contrast, a stray static charge or wayward magnetic field can wreak havoc on the media used to store digital sources."[29] To survive, digital media need vigilant tending. Benign neglect will not do.

But we aren't sitting by as our disks darken, and therein lies another nightmare: like men adrift in lifeboats, we will be surrounded by undrinkable flows. Digital information is not only inherently fragile, but delivered to us in forms that fast become obsolete as we bullishly upgrade to better machinery. Remember floppy disks and cassette tapes? "Any file stored more than six to eight years ago and not transferred to something more modern in the meantime, is on its way to doom," remarks technology writer James Fallows.[30] So like neo-medievalists, we are fated to copy, converting what we can into the language of the day, so it can be translated into a future technological tongue. Or we must build chameleonlike machines that can read old techno-speak. In a project that chief US archivist Allen Weinstein likens to the 1960s race to the moon, the federal National Archives and Records Administration is building a vast system that, if all goes well, will read the more than 16,000 software formats used in the estimated 347 petabytes of electronic data in its files by 2022.[31] Not all presidential e-mails will be saved, but millions will be, each one copied and saved and converted and translated as we battle to preserve what's essentially ephemeral. The hard-won result will be pockmarks of informational scarcity amid a shape-shifting abundance of information that will be hard, at best, to navigate or decipher. Even to catalog these slippery surfeits may be impossible, as a pioneering sixteen-month archival experiment led by the Library of Congress demonstrated.

The test seemed fairly simple, a bit akin to the children's game of "pass the egg." In 2004, the library sent four universities a relatively small, twelve-gigabyte collection of September 11–related materials, then the institutions exchanged the archive among themselves in order to see if they could "read" each other's versions.[32] Glitches were expected. "Ingesting" data is a lot tougher for an institution than just receiving e-mail, as new material must be formatted to the recipient's standards. The participating librarians and archivists, however, were stunned by the Babelesque result. Transferring the Linux-based collection first onto Windows altered many of the trove's fifty-seven thousand file names. "It was a big mess," says Martha Anderson, acting director of the Library's Digital Information Infrastructure and Preservation Program. "We made a false move, thinking very confidently that we knew what we were doing." Next, a kind of digital catalog of the contents in triplicate showed three different tallies of files in the collection, making it impossible to know if a recipient had gotten everything. The only accurate "metadata" turned up on a Library-designed format that worked in all-or-nothing style, so that if a file was corrupted, then the whole archive had to be resent. In the end, the sobering experiment showed digital preservation to be a messy game that we don't yet know how to play. "Ten years after the Web turned every institution into an accidental publisher, the simple difficulties of long-term storage are turning them into accidental archivists as well," wrote Clay Shirky in the final report on the experiment. Keeping stores of information may be the least of our challenges. Understanding what we've saved will be a far trickier task. Shirky, who teaches at New York University's Interactive Telecommunications Program, likens the problem to burning the card catalog at a major library; the books may as well be lost. "The real problem is findability," reflected Shirky more than a year after the test. "If you have no idea of what's there, it may as well be raw disk drive space. You can go through with forensic tools, but the amount of work required to find what you need would take more than a lifetime."

But have we ever inherited anything more than "fragments of a vanished whole," as archivist Laura Millar observes?[33] We have only a few of the greatest Greek tragedies and mere residue from the pyramids. The past is an imperfect pastiche of broken lace, the raw material of pattern making and interpretation. Moreover, no fragment can ever offer us the true essence of the original, as much as we yearn for an authentic glimpse of what's gone, writes Jacques Derrida in *Archive Fever*.[34] More than ever, we suffer from *mal d'archive*, "an irrepressible desire to return to the origin, a homesickness," writes Derrida, noting that the word itself comes from the Greek *Arkhe* for "commencement" and "commandment." Since his slim book derives from a 1994 lecture on memory given at Freud's house in London, Derrida naturally raises the ghost of the master. He recalls a 1907 essay by Freud about a fictional search for a perfect trace of the past. In the 1903 novel *Gradiva* by Wilhelm Jensen, a young archaeologist sees an ancient stone image of a young Pompeiian girl, Gradiva, and visits the lost city to travel in her footsteps. Ultimately, he winds up rekindling his childhood love for a woman vacationing in Pompeii with her zoologist father. Freud, who had a plaster cast of the Gradiva fragment in his consulting room and compared psychoanalysis to an archaeology of the mind, concluded that the young man's fetish for the image was his long-forgotten love for the girl. The story is a small reminder of how we're often too busy looking for traces of what's gone to notice that the past pushes into the present in different guises, as Arendt understood. "The question of the archive is not, we repeat, a question of the past," asserts Derrida. "It is the question of the future itself, the question of response, of a promise and of a responsibility for tomorrow."[35]

To preserve is not to recapture what's past but rather to change what's saved, at the very least by shifting the context to the new world of the present. "The paradox of preservation is that it is impossible to keep things the same forever," writes preservation scholar Michèle Cloonan.[36] "Even if an object survives untouched, it will have changed just by virtue of aging or by a change in its surroundings." The same holds true for cultural memory. Most countries commemorate only the very distant past or events of the past two hundred years, essentially relegating "stretches of history . . . to social oblivion," observes Eviatar Zerubavel. His study of "time maps" reveals that only 9 out of 191 countries formally mark on their national calendars any events from 680 to 1492.[37] And in a marvelous sense, the fate of our earthly possessions—our data, monuments, calendars—mirrors the biology of human memory, since to remember is to literally *re-create*. We think of memories as snapshots that we can pull out of the album of the mind, but remembering actually entails *reconstructing* a core memory "with subtractions, additions, elaborations, and distortions," notes Eric Kandel.[38] The "massive interconnectedness" of our memory stores can't help but shape our re-collecting, according to McGaugh.[39] The past is more of an unfinished business than we might at first imagine.

Can we face up to the age-old task of trying to learn from the past at a time when the raw material of knowledge is both stunningly fragile and stupendously abundant? Perhaps we shouldn't worry too much about preservation in the classic sense, suggests Umberto Eco in his essay "Living in the New Middle Ages."[40] In our time of "permanent transition," he writes, "the problem will not so much be that of pre-serving the past scientifically as of developing hypotheses for the exploitation of disorder, entering into the logic of conflictuality." The "disordered preservation" of the medieval era "was an immense work of bricolage, balanced among nostalgia, hope, and despair," he con-cludes. Eco is not counseling us to give up on the past. Not at all. Rather, he is calling on us to sift, to choose, to wrestle with the mak-ings of knowledge past and present all around us. We cannot do this, Cloonan warns, by keeping a "myopic focus" on the technology of preservation: the standards, formats, and systems that create vast dig-ital cupboards with cyberspace locks.[41] Nor will we learn from the past by indulging the impulse to save all, an increasingly popular collective and individual strategy for coping with data floods. The cost of a giga-byte of computer memory has dropped from ten million dollars in 1956 to a little more than thirteen dollars in 2000.[42] So we figure, why not save all, including the vast minutia of our lives? Leading computer sci-entist Gordon Bell calls the fruits of his multiyear "lifelogging" effort a "surrogate brain." Yet his personal trove of 101,000 e-mails, 15,000 documents, 99,000 Web pages, and 44,000 photographs is so big that he has trouble finding anything. "I either get nothing or I get too much!" he mutters while trying to find an article to share with a visiting reporter. Cyber-archaeologists and "software dumpster divers" will be the heroes of tomorrow, predict science fiction writer Vernor Vinge and others.[43] But will frenetically squirreling away as much as possible of our digi-bundance help us make sense of it? Remember, the Library of Alexandria also was a think tank, where scholars "mapped and measured the earth and the stars [and] invented modern scholarship," notes Alexander Stille in his elegant book *The Future of the Past*.[44] Recall that the thorny skill of *selection* was the foremost mandate of modern history's greatest collectors. Embracing and rejecting, sifting and culling —that is what we as "re-collectors" were born to do. Amassing towering alternate universes of saved experience marks the abdication of our own splendid, multifaceted powers of remembering—and forgetting. To ensure a future for our past and exploit the disorder of our day, we must be able to think for ourselves. Will we?

Around 1970, Walter Mischel launched one of the most famous experiments in modern psychology. He put a series of four-year-olds in a room with a bell and a marshmallow, then asked each if he or she would like a "small" or "large" reward. Most, if not all, opted for the bigger treat. But then the catch was revealed: only if the child waited until the experimenter returned, would he receive *two* marshmallows. If the child couldn't wait, he could ring the bell, and the adult would allow him to eat the first treat. While waiting, some children covered their eyes with their hands or arms, squirmed, talked to themselves, sang, created games with their hands and feet, and even tried to sleep.[45] "Seconds become minutes, not knowing when exactly the experimenter will come back, and the child feels growing frustration and temptation to ring the bell and take the immediately available treat," wrote the Viennese-born psychologist.[46] Some children rang the bell in a minute, others toughed it out for up to twenty minutes before the researcher returned.

For decades, Mischel and his colleagues at Stanford and then Columbia universities repeated the experiment, sometimes with pretzel sticks or toys, or after giving the children a helpful tip on waiting. And then, they tracked many of the preschoolers as they grew. Those who had waited longer for a treat became more socially competent, resilient, articulate, attentive, reasoning teens who scored substantially higher on the SATs and got higher educational degrees. Kids who opted for instant gratification were more likely to become bullies, get lower parent and teacher evaluations as teens, and abuse drugs as adults. Even seconds' difference in wait times made a difference in the trajectory of a life. How could such a breathlessly simple experiment predict so much? Most importantly, what can this Delphic nursery game tell us about sense-making in a culture of distraction? To begin connecting the dots, consider the worrisome state of US higher learning, by many measures a national asset in decline. For as Arendt notes, our capacity for thought, "unlike the world and the culture into which we are born, cannot be inherited and handed down by tradition."[47] Each of us "must discover and ploddingly pave anew the path of thought." The marshmallow test gives clues, for those who look closely, to understanding the mental muscle we need to bring to the lifelong enterprise of path making.

"Although Americans today are more highly educated than ever before, they are not necessarily better educated," writes Lisa Tsui, a researcher at the Urban Institute.[48] At about 40 percent, we rank a close second to Canada in a comparison of the proportion of adults ages thirty-five to sixty-four who hold college degrees in nearly thirty developed countries. But we place seventh in similar international rankings of graduates ages twenty-five to thirty-four.[49] At most, 65 percent of students return for a second year of community college or get a four-year degree within six years of enrolling. Moreover, many have unrealistic expectations. Ninety percent of college freshmen expect Bs or better although most study only about fourteen hours a week, half of what faculty expect, according to the National Survey of Student Engagement. Nearly a third of first-year students say they work "just hard enough to get by," while 20 percent of seniors report frequently coming to class unprepared.[50] Students increasingly must deal with rising costs and a resulting need to work, yet many simply are not prepared for the academic challenge: only about a quarter of high school graduates are college-ready in the four main subject areas of English, reading, math and science.[51] While many of these challenges aren't new, nonetheless, they are collectively disturbing. Close to half of students enter college needing remedial help,[52] and many exit still needing remedial help, while making mediocre gains in the one skill that, as Harvard president Derek Bok notes, almost all educators agree is the main purpose of an undergraduate education: critical thinking. "Even four years of college only brings traditional-age college students to a very low level of critical thinking and judgment," says Patricia King, an educational psychologist who has spent three decades studying higher-order thinking. "They're making what we call quasi-reflective judgments."[53]

Like its illusive sister concept—information literacy—critical thinking is defined in numerous ways. Bok describes critical thinking as "habits of disciplined common sense" that include reasoning, problem solving, judgment, and good research skills.[54] Others consider methods of logic or even visual observation skills to be at the heart of critical thought, a field largely inspired by the influential writings of John Dewey. But however defined, students seem to be making at most lukewarm and perhaps declining progress in this area of paramount importance. On average, seniors who rank in the middle of their freshman cohort in this ability leave college with scores equal to no better than the seventieth percentile of their freshmen cohorts, with most gains made by the end of sophomore year, report Ernest Pascarella and Patrick Terenzini in a massive meta-analysis of 1990s data on the impact of college. A decade earlier, students scored nearly twice the gains, report Pascarella and Terenzini, cautioning, however, that their previous analysis may have been an overestimate.[55] Students also get better at "reflective judgment," the kind of thought needed to wrestle with ill-structured issues, such as gun control. Yet most college students graduate still believing that knowledge is just a matter of opinion, in which evidence plays little role, notes Patricia King.[56] Children and other prereflective thinkers insist that seeing is believing and all problems have definite answers, then gradually begin to understand that uncertainty is an ally of good thinking and knowledge is an evidence-based construction. King's work shows that during college, students trade "ignorant certainty" for something akin to "intelligent confusion," emerging into the world "naïve relativists."[57] The question is, in a land of distraction, can they become truly reflective citizens—"convinced that ill-structured problems are susceptible to reasoned arguments based on evidence," not just opinion?

Walter Mischel did not dangle treats before young noses simply to test preschooler patience. He set out forty years ago to untangle the inner workings of how people attain their goals, and especially how they manage to keep pursuing rewards that don't come quickly or easily. "What makes it possible for some people to give up their addictions, to resist the temptations that threaten their cherished values and goals, to persist in the effort, to maintain their relationships, to overcome the more selfish motivation and take account of other people?" he asked.[58] At the time he began his research, psychology was still under the spell of behaviorism, which held that people could be largely understood by studying their visible actions. Mischel, in contrast, was fascinated by what he could not see: the thoughts that shape the course of our lives. In his research on "delay of gratification," he found that if he suggested that the preschoolers think about how "yummy and chewy" the marshmallows were, they broke down in five minutes. But if they coolly thought of the marshmallows as "white puffy clouds" or pretzels as "little brown logs," they could wait an average thirteen minutes. The children's cognitive control so influenced the outcome that they could wait almost eighteen minutes simply by pretending that the treat before them was a picture. (Asked later, one child explained, "[Y]ou can't eat a picture.") In contrast, a preschooler shown only a *picture* of a reward but asked to imagine that it was real lasted less than six minutes. Imagining the treat in all its proximate glory triggers our emotional, "hot" selves, while thinking about the reward in cool, distancing, contemplative terms makes waiting easier, Mischel discovered.[59] A few hundred marshmallows later, Mischel was beginning to uncover the mechanics of self-control, a concept central to humanity in William James's view yet largely set aside as a subject of study for much of the twentieth century. The ancient Greeks had a term for impulsivity: *akrasia*, a deficiency of the will.

Does willpower have a place in twenty-first century learning and growth? Certainly, the notion smacks of outdated, if not downright prudish, notions of rule-bound discipline. Yet in reality, self-control is a complex and fascinating concept that is central to fostering the reflective thinking skills and deep engagement in learning that are so needed, individually and collectively, in the digital age. Willpower is a part of self-regulation, or the ability to control body and mind and transform "the inner animal nature into a civilized human being."[60] Self-regulation can be automatic, such as when body temperature is maintained, or more notably, it can involve a conscious and "delicate balancing" between short-term interests and long-term goals. Without self-control, we can have the strongest of motivations and set the highest of goals, yet we will invariably get sidetracked in pursuing our aims by the distractions, temptations, and obstacles of life—from channel surfing and chocolate cake to problems so tricky that we give up doing our best. When Martin Seligman and Angela Lee Duckworth studied 164 eighth-graders over a school year, their scores on measures of self-control proved *twice* as predictive as IQ of final grades, selection into a competitive high school, hours spent doing homework, hours *not* spent watching television, and time of day that homework was begun. The two psychologists even gave the students a big-kid version of the marshmallow test, handing them a dollar in an envelope that they could take immediately or trade in a week later for two dollars. Self-control, in essence, predicted the model middle schooler, just as a few marshmallows could give tantalizing clues to a preschooler's future. Yet "for every article on academic achievement and self-discipline in the PsycInfo database, there are more than ten articles on academic achievement and intelligence," conclude Seligman and Duckworth. "Underachievement among American youth is often blamed on inadequate teachers, boring textbooks and large class sizes. We suggest another reason for students falling short of their intellectual potential: their failure to exercise self-discipline."[61]

But what puts the power in willpower? *Attentional control* is the driver behind willpower and the means of successfully resisting gratification. Recall that focusing on the yummy-ness of the treat sabotages the youngster's efforts, unless they distance themselves from its desirable qualities by pretending it is a picture or a cloud. Self-distraction, via focusing on a fun memory or a small toy, also helped children wait more than ten minutes—as long as if the reward was out of sight. In effect, the children were using their powers of focus to cause the temptation to disappear. All in all, the most patient children often were junior attentional athletes, those who "seem to focus briefly on the hot features in the situation to sustain motivation but then quickly switch back to the cool features and self-distraction to avoid excessive arousal and frustration," note Mischel and colleague Ozlem Ayduk.[62] Here again we see how executive attention is the "conductor" of the orchestra of you—helping you hold your goals and perceptions in mind, tamping down the drumbeat of your short-term wishes and unthinking emotional responses, and canvassing your cognitive and external worlds for help in attaining those aims and avoiding obstacles to doing so. Far more than a "just say no" skill, self-control also gives us the gift of "what-if," an inner life that offers us the chance to mentally test out the future without suffering "real world consequences for one's mistakes," notes attention-deficit/hyperactivity disorder (ADHD) researcher Russell Barkley.[63] In other words, self-control is the essence of looking—literally and cognitively—before you leap.

Our extraordinary capacities for self-control begin essentially with a baby's gaze. Don't be fooled by the seeming helplessness of a three-month-old, assert Michael Posner and child psychologist Mary K. Rothbart in *Educating the Human Brain*.[64] Infants may not walk or talk, yet already they are learning to control their gaze and their attention, abilities that allow them to "enter into their own education" by interacting with others and learning about their environment. With help from their caregivers, they learn next how to begin to control their emotions, first by turning away from sources of distress. In one study, eight- to ten-month-old infants who were more capable of focused play also showed less anger and discomfort when exposed to a fifteen-minute cavalcade of mostly "aversive stimuli," including a clown mask, the touch of an ice cube, a taste of unsweetened lemonade, and even a researcher grabbing a toy from them.[65] Last to develop over the course of childhood is the executive attention system, centered in a small, banana-shaped front brain region called the anterior cingulate, whose very anatomy illustrates the fact that self-control is a constant balancing act between thought and emotion. The ventral—or back half—of the anterior cingulate helps regulate emotion, while the dorsal, front half manages the control of cognition, and generally one inhibits the work of the other.[66] So when a child coolly turns a marshmallow into a cloud through his powers of focus, he's effectively ramping up the cognitive half and tamping down the emotional side of his anterior cingulate. Of course, willpower is not purely rational: moderate doses of both positive and negative emotion boost cognition. A quick peek, not a long, lustful look, at the treat is good ammunition for the skillful child-in-waiting. But mostly, we think best when we are not too carried away by emotion. Overall, fueling self-management is yet another way that our attentional skills allow us to live in an astoundingly complex world without falling constant prey to distraction. Moreover, it's fitting that the marshmallow test highlights the potential suppleness of our attentional powers, since Mischel is also renowned for his work on the flexibility of personality expression. Shy people can be extroverted and optimists can get depressed, he has long argued, just as we can turn marshmallows into clouds.

Are we missing a key to learning? We need self-control to finish a project, persevere through tedious or seemingly indecipherable bits of life, hold our temper, plan for tomorrow, and turn off the television to meet a baby's gaze. Minute by minute, self-control blossoms into broader forms of focus and persistence called "engagement" that fuel both academic achievement and depth of thought. Students who study longer and get more deeply involved in their courses, and their reading and writing, report making greater gains in critical thinking and intellectual development in college, numerous studies show.[67] This is both common sense and somehow surprising. Surely, the kids who take detailed notes and speak up in class, live in the library, and devour the list of "additional readings" naturally learn more, but they also wind up becoming better *thinkers*. Moreover, self-controlled students manage their efforts at learning not only by suppressing distractions but by constantly evaluating their own thought and progress.[68] Just as babies first learn through visual self-control, so successful students from kindergarten onward *shape* their educations, with the help of good teachers. No wonder the American Philosophical Association in part defines critical thinking as "purposeful, *self-regulatory* judgment" (italics mine), and Derek Bok argues that "*disciplined* common sense" (italics mine) drives critical thinking.[69] Certainly, we don't want to nurture willpower to the point of rigidity, cautions Mischel. "Excessively postponing gratification can become an unwise—and even stifling—joyless choice," he writes, "but unless people develop the competencies to sustain delay and continue to exercise their will when they want and need to do so, the choice itself is lost."[70]

Perhaps attention is the true missing key to better learning. Without the powers of focus, awareness, and judgment that *fuel* self-control, we cannot fend off distractions, set goals, manage a complex, changing environment, and ultimately shape the trajectory of our lives. In a culture of distraction, could the erosion of our skills of attention be linked to some of the alarming ills weakening our higher education system? Certainly, many college students are focused, disciplined, achievement-oriented, and committed. Yet we as individuals and as a society should be concerned by the fact that many campuses are plagued by high dropout rates, lukewarm gains in critical thinking and reflective judgment, and low engagement with the life of the mind. A disturbing thirty-five-year longitudinal study by Alexander Astin reports that 80 percent of incoming college students in the 1960s called "developing a meaningful philosophy of life" their most essential goal, while 45 percent chose "being very well off" as their primary aim. Today, their values have shifted markedly. Nearly 74 percent of freshmen rate being wealthy as essential, while 43 percent choose life philosophy as most important.[71] Surely, making money is a worthy and thoroughly American goal that takes self-discipline. Further, the 1960s may have been a particularly idealistic time. And yet something seems lost when students don't place high value in wrestling cogently with the meaning of their time on earth. In a sharply critical report on campus life based on five years of studies at the Harvard Graduate School of Education, educators Arthur Levine and Jeanette Cureton conclude that many of today's students are career-oriented yet academically and socially unprepared.[72] Could a culture of distraction be hindering our eternal human struggle to shape our fates?

The renowned ADHD researcher Russell Barkley argues that the capacity for executive function and self-control may have evolved in humans from two crucial social needs—exchanging goods and services with those outside the family group and learning through imitation.[73] Such achievements essentially involve forms of waiting and remembering, in other words, the "mental spreadsheet" skills governed by the executive attention network and lacking to varying degrees in people with attention deficit disorders. In periods of alternating plenty and famine, it would behoove lucky hunters or farmers to share extra supplies in exchange for later shares in another's bounty. Similarly, our ability to sustain the memory of a skill, such as tool-making, over time and distance and then replicate the behavior for one's own purposes, is a crucial and uniquely human capacity. Those with ADHD, however, suffer from forgetfulness and impulsivity, traits that impair their ability to shape the course of their lives. "The child with ADHD will be more under the control of external events than of mental representations about time and the future, under the influence of others rather than acting to control the self, pursuing immediate gratification over deferred gratification and under the influence of the temporal now more than of the probable social futures that lie ahead," asserts Barkley, concluding that ADHD is a disorder of "attention to the future and what one needs to do to prepare for its arrival." It is, he says, a "disorder of time."

Averting a dark age isn't simply a matter of saving disks and instilling discipline. We surely need to wisely preserve our digital heritage so that we don't become either informational amnesiacs or mindless hoarders of life's minutia. We need a past to learn from. At the same time, we should endeavor to value self-control as integral to many, if not most, functions of life and a key to deeply engaged, critical learning.

But this is not the full story. Only a society rich in memory and reflection can hope to build a culture of creativity, vision, and care. Cultivated stores of memory and abundant capacities for thought— individually and collectively—carry us forward, giving us the means to face coming challenges and to produce the next great artistic and literary masterpieces. From memory, we gain the potential to layer our present with the teachings of experience. In reflection, we find what Virginia Woolf literally and metaphorically called a "room of one's own," a space for both consideration and creativity.

And how are we faring at that mutable place where, as Hannah Arendt describes, the past meets the future in an eternal struggle? Heads down, we are allowing ourselves to be ever-more entranced by the unsifted trivia of life. With splintered focus, we're cultivating a culture of distraction and detachment. We are eroding attention—the most crucial building block of wisdom, memory, and ultimately the key to societal progress. In attention, we find the powers of selection and focus we so badly need in order to carve knowledge from the vast, shifting, and ebbing oceans of information that surround us. From attention, we glean the will and the tenacity to create lives of meaning and cultures marked by reason and vision. In cultivating distraction, we cannot perceptively look back or see ahead; we are defeated in the battle to shape our future.

And yet, a *renaissance of attention* may be at hand. An antidote to our epidemic distraction lies in a set of astonishing discoveries: attention can be understood, strengthened, and taught. However we may define progress now or in future, there is a spark of hopefulness in that.

Chapter Nine

THE GIFT OF ATTENTION

A Renaissance at Hand

A chill December drizzle cascaded down my borrowed poncho and soaked my jeans, and raindrops trickled off my nose. I was trudging through the Oregon woods with Mike Posner, unarguably the greatest attention scientist of our time. My boots sank into oozing mud as I strained to follow Posner, a lanky septuagenarian, along a winding creek-side trail close to Alderwood, his weekend home. Seattle-born and an Oregonian for decades, he's unruffled by unending gray and rain. Striding forward, hardly pausing, he sporadically launched observations on the landscape over his shoulder, his unforgettably deep, rumbling, and yet somehow comforting voice rising easily above the wind and rain. He talked of the fish returning to the stream and the fir trees he's planted along the trail. Damaged by animals or blight, a number of the pint-sized evergreens weren't thriving, but again Posner didn't comment. He's someone who more often sees distant possibilities than at-hand impediments; a man who told his wife, Sharon, a half-century ago that what he really wanted to do was to understand how the brain works. Attention became the key to this quest, for he who cracks the riddle of attention holds one of the most crucial secrets to fathoming the human mind. By giving us the framework and tools for decoding the enigma of attention, Posner has offered us the means for both understanding and shaping ourselves. He holds clues to combating a darkening time.

Here at the end of my journey, I am looking for the light. I have stood vigil lakeside with a grieving father and held hands with a helpful robot, exploring our land of distraction. I have strolled through Brooklyn with an obstreperous cartographer, tasted a car-culture pizza, tested my multitasking prowess, marveled at the charisma of a long-deceased "auto-icon," cradled an antique Bible in my palm, and learned that attention is becoming a casualty of our newly virtual, split-screen, and nomadic world. In adapting to these stunning new experiences of space, time, and place, we are allowing the pillars of our humanly attention to suffer. Focus slides into diffusion, judgment becomes skimming, and awareness slips into detachment. The antidote, of course, is attention. But to cultivate a *renaissance of attention*, we must understand its workings and have the patience and wisdom to envision its possibilities. "Everyone knows what attention is," wrote William James in his 1890 masterwork *Principles of Psychology*. "It is the taking possession by the mind, in clear and vivid form, of one out of what seem several simultaneously possible objects of trains of thoughts."[1] James gives us a snapshot of the end point of focus, the moment when we capture and hold something—a thought, a taste, a memory—in mind. Posner often opens lectures with this quote, then notes that nowhere does this catchy phrase mention the mechanisms of the mind. No one a century ago could peer into the pulsing brain, eavesdrop on synaptic conversations, or trace the complex lines of cognitive coordination needed to entrap, possess, and perceive a bit of the world. Now we can, and even more astonishingly, we are well on our way to understanding one of the brain's most miraculous feats: the multifaceted networks of attention, a cognitive tour de force whose mechanisms may unearth the nature of consciousness itself. On the trail of decoding attention, Posner is invariably two steps ahead. He's the opposite of a magician, who spins illusion from our propensity for distraction and our inability to fully perceive the world. Instead, he's spent a lifetime making the invisible *visible* and bringing the mysteries of the brain to light—so that we can better chart our own lives. For not only can attention be understood, it can be shaped, strengthened, and trained. Reflecting on the majesty of these discoveries as I tromped through the forest, I lost my footing and stumbled. Looking up, I saw Posner forging ahead.

If our three networks of attention—orienting, alerting, and the executive—are comparable to an organ system like digestion, then *orienting* is akin to a cognitive mouth, a gateway to our perception, the scout. Orienting is focus deluxe, the most nimble attentional offspring, the acrobat that allows us to perceive something new in our environment, swivel our attention to it, and determine its importance. If you are lost in the forest, orienting is a very good skill to have. When explaining attention to her five-year-old, cognitive neuroscientist Amishi Jha uses a perhaps more elegant analogy, comparing orienting to a flashlight of the mind. Either way, consider orienting our wayfarer, a means from infancy onward to understanding life and to nurturing the deep interconnectedness that we need to thrive. And while we orient to noises, smells, or touch, this form of attention is best understood visually, for we are, as we've witnessed, a species ruled by sight. At birth, only the deepest of the visual cortex's six layers contain neurons. Then, during the first year of life, the cortex develops from the inside out, with the maturation of the deeper, more primitive sections followed by the newcomer neighborhoods in evolutionary terms.[2] The hidden abilities of these older brain pathways account for "blindsight," the ability of some people to "see" a little in an area of their visual field that is blind. They report seeing nothing to their right, for instance, yet can reach out and accurately locate a light or object held up in this dark, sightless area. This is because spatial location is determined by subcortical brain areas that work with parietal lobes to help us navigate the world spatially.[3] Blindsight is the opposite of visual neglect—the syndrome caused by parietal damage that causes a lack of awareness on one, usually the left, side of the world in someone with normal sight.

By four to six months, babies are "looking machines," marvels Posner, casting their tiny spotlights of attention on one after another item of interest. Gradually, two separate parts of the infant brain's parietal lobe—an area at the crown of the head closely related to eye movements and other sensory processing—learn to work with other brain areas to allow the child to focus. At first, a baby orients willy-nilly, to the bright, shiny, or noisy, but gradually, she can focus on nearly anything she chooses.[4] In neurological circles, this is as exciting as a baby's first steps. Early on in this intense learning period, just as month-old infants are beginning to master the idea of focusing, but before they've learned how to turn away, they go through a months-long period of "obligatory looking" that lays the groundwork for a strong bond with a caregiver.[5] They can't seem to turn from stark patterns such as checkerboards and are especially prone to staring deeply into another's eyes, like swooning little lovers. Later, new things will almost invariably catch their eye, a lasting bias to novelty that helps humans efficiently search—and thus survive—their changing environment. But the young human, an ever-curious discoverer, is simultaneously built to *share* this burgeoning ability to explore. Like baby monkeys riding piggyback on their mother to see the world, healthy human six-month-olds can look where a caregiver looks, then by eighteen months, understand when they are sharing a moment of focus. This is *joint attention*, a seemingly simple but extraordinarily rich moment of first connection, courtesy of the developing orienting network. While still incompletely understood, its profound importance is nonetheless appreciated by scientists. Joint attention is a "meeting of minds" that is critical to social, language, and cognitive development, with a role in developing early verbal communications that "is hard to overestimate," writes philosopher Naomi Eilan.[6] One of the first signs of autism is a deficiency in joint attention. Autistic children miss others' "bids" for sharing attention and don't initiate shared focus, so become marooned socially, frustrated with their solitude, yet unable to connect to others.[7] Orienting, then, is much more than a scout, it's our bridge to one another. When we opt again and again for the new, new thing, and grow sated with snippets and glimpses bred in the seductive optionality of the virtual realm, aren't we risking a kind of social autism? When we relate via bits and pieces of each other gleaned from surveillance, aren't we corroding the power of this rich attentional skill? Lose the will to focus deeply, to point the compass of our lives firmly in one another's direction, and we become islands.

Back home, Posner settled into a favorite chair, with a mug of coffee and his chocolate-and-black miniature schnauzer, Annie, by his side. Rain pattered on the cone-shaped roof of the simple, octagonal wooden yurt, as I peppered him with questions about what is perhaps the greatest detective story of our time: understanding how the brain— and especially attention—works. Decoding this mystery begins with a question of real estate. Can thinking be traced simply to the astonishing mechanics of the brain, or is there something more? Is consciousness soulful? The view espoused by seventeenth-century philosopher René Descartes that brain and mind are sharply separate or "dual" still shadowed the psychology field that Posner joined in the late 1950s. Nineteenth-century French surgeon Paul Broca and others had roughly mapped brain areas responsible for activities such as movement and speech, a discovery taken to extremes by phrenologists of the day who sought to connect intellect to cranial and even facial bumps. But for the most part, psychologists, who study the mind, and neuroscientists, who study the brain, were not on close terms when Posner began parsing the mechanics of cognition in order to realize his dream of deciphering the engine of the brain. At the time, attention was a conceptual will-o'-the-wisp, thought to be a kind of cognitive filter or perhaps a pantry of skills. It was hardly the stuff that could be captured and even filmed by a young scientist unwilling to kowtow to scientific fads or naysayers.

Early on, Posner ran a simple experiment that profoundly affected his thinking. He found that people take eighty milliseconds longer to recognize that a pair of upper- and lower-case letters, for example, "Aa," are the same letter than if they are shown two of the same case, "aa" or "AA." Those tested had spent a lifetime looking at this letter, yet they stumbled over this tiny variation. Why? At the time, Posner didn't know that right-hemispheric visual brain regions interfere a bit with efforts by the left-hemisphere's visual word form area to categorize this alphabetic tidbit. But he did correctly surmise that widely separate areas were firing in response to differing case combinations—so the phrenologists weren't all wrong in their efforts to connect thinking to brain regions.

His next area of research marked a milestone in the history of experimental psychology: Posner created an ingenious way to measure attention—a seemingly impossible feat. We all know what it feels like to pay attention, to swivel our focus or give a problem our full concentration. But how do we *measure* this essentially invisible process? "We can measure the genome, molecules, reaction times, behavior, but how do we engage in scientific research about something that is a psychological construct?" asks neuroscientist Bruce McCandliss. "That's the genius of Posner's cueing test. He came up with a clever paradigm for measuring an invisible ghost inside the head."[8] The task Posner created is simple. He asked people to press a button when they saw a stimulus, such as a letter or asterisk, on a screen, then measured how much quicker they reacted when warned where on the screen the stimulus would appear. The key is that people weren't allowed to move their eyes but rather only absorbed the warning by *orienting* their attention to it covertly. Presto. In what has come to be called the "Posner test," Posner was able to measure the imprint of attention, like a child's invisible ink coming to life before a flame.

Throughout his early career, Posner hunted down the nature and genesis of attention. He crafted ingeniously simple time-based experiments, probed stroke victims' deficits, and finally shocked colleagues in the mid-1980s by gambling that the new field of neuroimaging—scorned as the "new phrenology"—could marry studies of mind and brain. Working with Marcus Raichle at the Washington University in St. Louis School of Medicine, Posner helped put the nascent field of cognitive neuroscience on the map, unveiling in the process what he'd surmised all along—the networked nature of attention. In a visionary paper now largely forgotten, he had outlined the tripartite workings of attention—in 1971.[9]

Orienting is eager, but *executive* attention has swagger. The executive network is our troubleshooter, a sheriff who moonlights as the judge, with a long reach and a heady power that is, alas, easily corrupted. The modus operandi of this network, with its mastermind, the anterior cingulate, remains a bit of a mystery. Yet without this capacity, the brain is a lawless patch of cerebral country. Executive attention is not only at the heart of self-control but central to a spectrum of crucial capabilities from planning to meta-cognition, that is, understanding the workings of the mind. Encounter a snake on the trail? Write your will? Multitask? Quit smoking? Almost all new, difficult, dangerous, or complex challenges or occurrences prompt a 911 to executive attention, whose résumé touts "selection" and "control," with a specialty in "conflict resolution" and "catching mistakes." Executive attention operates "top-down," meaning higher-order brain regions draw on sensory data to impose a framework of interpretation and volition on our thoughts and actions. Bottom-up processing, in contrast, carries raw perceptions up a ladder of cognition, from lower to higher echelons of the brain. We constantly operate via both circuits, of course, yet much of voluntary consciousness is top-down, hence we think we are in control even when depending on fairly skewed versions of reality. Hypnosis seems to work by essentially dimming that mastermind of executive attention, the anterior cingulate, leaving you susceptible to highly irrational suggestions, as neuroscientist Amir Raz has discovered.[10] The power that so spooked the Victorians is now wielded by a new breed of researcher probing what Raz calls "atypical attention," from trances to comas or meditative states, to plumb the workings of the brain. Raz got a head start in this pursuit. A former professional magician, he became an attention researcher to track the slippery beast he'd manipulated so well on the stage.

To study executive focus under hypnosis, Raz used neuroimaging and the best-loved tool of attention scientists worldwide—the Stroop test. Invented in 1933 by a young Tennessee farmer-turned-scholar named J. Ridley Stroop for his dissertation, the test is clever and simple. Shown an all-caps color noun such as "BLUE" that's shaded green or red and asked to name the ink color used, we invariably have to think hard, essentially because reading is so automatic that we feel like mistakenly blurting out "blue." The Stroop effect nimbly tests a core function of executive attention: conflict resolution, a capacity integral to judgment and decision making. After being hypnotized by Raz and told that they would be tested on "meaningless symbols," sixteen people returned days later to the lab and—while not hypnotized— underwent a Stroop test in a brain scanner, with those previously most easily hypnotized showing little hesitation, conflict, or cingulate activity. No one fully understands how our executive network works, but the answer may stem in part from the long grasp of the anterior cingulate, an ancient workhorse of the mammalian brain that has evolved significantly in humans and primates. Deep within the cingulate and insula of humans and great apes is a type of long-fibered cell that is thought to coordinate these regions with other cognitive and emotional brain systems.[11] The far reach of executive attention, its importance to much of what we deem human, plus its relatively late development conspire to make this network prone to misfire. Schizophrenia, attention deficit, borderline personality, and a host of other disorders are linked to breakdowns in executive attention.[12] Day to day, we know that the cultivation of control, judgment, and planning skills is not easy for anyone. Yet we cannot reshape a culture built upon split focus and info-skimming without harnessing the powers of this network. Executive attention is a key to turning back a tide of distraction.

My weekend at the yurt ended, I visited Posner's lab at the University of Oregon, where he arranged for a colleague's grandchildren to demonstrate a few simple attention tests that are helping to illuminate the fascinating riddle of how attention develops. Eighteen-month-old Lucy chewed on her finger and rubbed her eyes after a few minutes of orienting to cartoon animals flashing on alternating computer screens. Six-year-old Emma flew through several computer-based tasks, including a child-friendly, number-based version of the Stroop and a short exercise involving helping a cat move around a yard. Both were developed by Posner and colleague Mary Rothbart. Are some of us born attention-deficient and others highly focused? Or is there a time when children's attentional capacities can be strengthened, just as we can boost their math and reading skills with the right kinds of drill and practice? Can we fine-tune awareness in adulthood or battle the ravages of stroke through an attentional cure? These are the questions that have inspired the second half of Posner's career and his quarter-century partnership with Rothbart, who pioneered the study of temperament and effortful control.

Their collaboration began, in a sense, from an academic dare. Stung by a famous developmental psychologist's complaint that the lab-centric work of cognitive researchers tells little of real life, Posner invited Rothbart in 1980 to coauthor a paper on "The Development of Attentional Mechanisms."[13] It wasn't easy. While both are psychologists, they speak different professional dialects: Posner's early work centered on orienting attention without eye movements, for example, yet eye tracking is manna to baby watchers like Rothbart or Daniel Anderson, the television researcher. Still, over the years, Posner and Rothbart have hammered out numerous ways to study the attentional prowess of infants and children. In one experiment, six- to nine-month-old babies watching toys being added or subtracted from behind a screen look longer if the number revealed doesn't add up to what they'd seen placed there. Essentially, the babies recognized simple math errors—a sign that a prototype of the executive network is in place even in infancy.[14] Through batteries of tailored cognitive tests coupled with parent reports and electroencephalograms, Posner and Rothbart have helped show that our attention networks develop independently on a rhythm all their own. Orienting is largely in place by kindergarten, and the core skill of conflict resolution peaks by eight. Alerting plus some other aspects of executive function still develop into adulthood.[15] The magic of our attention networks unfolds over many years, shaped by both birth and background.

For a last clue in finding the genesis of attention, consider the role of genetics. Is our destiny etched in our twenty-five thousand genes or are we more a product of experience? This is a question I've long pondered as an adopted twin, raised with my sister yet ignorant in other ways of my heritage. The answer, of course, is that we are a murky blend of genetics and experience, no less in our attentional prowess than in other realms of life. As parents, "we are limited in the good we can do and the harm we can do, and maybe that's a good thing," muses Posner, who likes to recount a small but tantalizing finding from his latest longitudinal research into the development of attention. The protagonist is dopamine D4 receptor, a gene built to handle the interneuronal flow of the neurotransmitter dopamine, which modulates a slew of functions, including our perception of whether an experience is rewarding. One version of DRD4, called the 7-repeat, is common among people with attention-deficit/hyperactivity disorder and has been linked to sensation seeking.[16] After swabbing, testing, and monitoring a crop of seventy toddlers since infancy, Posner and Rothbart found that parenting sometimes casts a kind of swing vote of influence on a child's genetic makeup. If a two-year-old has the 7-repeat and is exposed to even mildly poor—read dictatorial—parenting, she is likely to be an impulsive, high-energy sensation seeker, ripe for later attention disorders. Without the 7-repeat, quality of parenting makes little difference. "The gene might be making it more possible for the environment to have its influence," says Posner. Genes aren't destiny, and dissecting their influence on behavior is extraordinarily complex. But just as varieties of brain lesions and little timed tests can shed light on the workings of the normal brain, so genetic variations can illuminate the anatomy of our attention networks—and eventually, the workings of the mind.

Breaking down the mysteries of life into comprehensible pieces—that's why Posner is a legend in his own time. A member of both the prestigious National Academy of Sciences and Institute of Medicine, he has been ranked by his peers as one of the one hundred greatest psychologists of the twentieth century.[17] His vision of attentional networks is not universally accepted, yet it remains incomparably influential, as does his brilliant body of work. Looking into the future from his Pacific coast yurt, Posner sat up, raising his gravelly voice. "It's possible that now you will be able to figure out how a particular gene on a particular node in a network is related to building a particular network, and then you'll be able to build a construction plan for the outline of the network," he said excitedly. "That's why this is interesting—because you can have that promise." It's his half-century dream becoming real—a blueprint for attention that begins to outline the architecture of the mind. A few months later, I met up with Posner again as he visited New York, where he was the founding director in 1998 of the Sackler Institute for Developmental Psychobiology at Cornell's Weill Medical College. He lectured at the American Museum of Natural History and Mount Sinai School of Medicine, then stayed on at the school for a takeout Chinese lunch with a few dozen researchers who eagerly solicited his advice on their work. One psychiatrist described her intricate efforts to study how emotion affects interruptions. Posner briefly made a suggestion that I could not hear, and the room grew quiet; the researcher was clearly thrilled. Later, she told me that he had suggested studying the influence of aspects of attention on emotion. He reversed the question, offering an exciting new perspective on her research. When Posner speaks, scientists take note. And perhaps we should listen, too. Could the explosion of research that he has helped spark inspire an attentional renaissance for us all? Armed with a burgeoning understanding of its mechanisms, we can begin to cultivate our potentially limitless powers of attention. I took a tentative step in this direction not long ago eight thousand feet up in the Colorado Rockies. But before heading to the high peaks, let's visit the third and most lowly network of attention: alerting, a foremost building block of mind.

Alerting is the gatekeeper network, the caretaker who turns the lights on and off and keeps the hearth fires burning in our cerebral house. Simply put, alerting is wakefulness, the cornerstone of sensitivity to our surroundings. It comes in many flavors, from a coma to a coffee buzz, and is as necessary to life as the air that we breathe.[18] Still, the study of alertness has long received short shrift in scientific circles, aside from a stream of studies focusing on how well and how long air traffic controllers, truck drivers, and other workers can stay awake. "I don't think people have realized how difficult and complicated the alerting process is," says Posner. "It's a very complex state." Consider what happens when adrenaline-fueled arousal systems go awry: patients in comas or slightly more wakeful vegetative states show no signs of awareness, although they eerily sometimes cry, grimace, or talk.[19] Glitches in how the neurotransmitter noradrenaline modulates the alerting process may account, as well, for some share in the spectrum of ADHD troubles, argue psychologist Jeffrey Halperin and others.[20] Even hour by hour, alertness ebbs and flows like an ocean tide. When we awaken, we suffer "sleep inertia," a twenty-minute stint of "dampened responsiveness."[21] Then, we grow more alert all day, until evening. This fluctuating "tonic" alertness is rooted in the brain's right hemisphere, while "phasic" alertness—the get-ready feeling triggered by a teacher's bark or a starting gun—is thought to be left-brained. In switching to the phasic ultra-on mode, our alerting network simultaneously quiets down other cognitive activity, partly with the help of the anterior cingulate. There is "crosstalk" between the attention networks, notes Raz, although they largely operate independently.[22] And so humble alerting hoists the flag of wakefulness and holds the keys to our castle-keep. It's our ticket to the beautiful world all around us. When we race breathlessly through life, detached from our surroundings and addicted to a kind of mindless mobility, we are short-circuiting this third power of attention. When we seek relationships with our inanimate machinery, we begin to deaden ourselves to the pinpricks and caresses of the real world.

Before leaving Oregon, I asked Posner a last question: what does he think of William James, the thinker who spearheaded the first modern teaching and research into psychology in this country a century ago? Lauded as a psychologist in his day despite his controversial interest in the paranormal, James shunned experiments and was largely a theorizer.[23] Yet in other ways, parallels to Posner are intriguing. Broad-brush, unpretentious men, they share a rare gift for distilling and dissecting big, thorny concepts, partly by anchoring them to the real world. Lucid writers and charismatic lecturers both, they dedicated their careers foremost to the development of human potential, in theory and day to day. Partly inspired by a domineering father and early paralyzing self-doubt, James fiercely supported the notion of free will and feverishly sought to inspire others to make the best of their lives. He encouraged and challenged his students at Harvard, holding office hours for years at home over dinner. "He had the rare ability to make people believe that they could find within themselves lush new corners in the gardens of their spirits," writes biographer Linda Simon.[24] In 1892, a young protégée wrote to James, who was abroad: "Cambridge without you is like toast unbuttered."[25] Posner, in turn, has spent decades investigating how to nurture attention in children and is famous for his generosity as a teacher: he shuttles visitors to and from the Eugene airport, invites students to supper, buys a platter if his postdoc's wife is a potter. But no, William James is not Posner's hero. He was a genius, said Posner, but he was really a philosopher, not a scientist, in the modern sense of the word. And on one essential point, the philosopher and the scientist certainly diverge. James thought that attention could not be highly trained "by any amount of drill or discipline."[26] Posner and others, nonetheless, are beginning to prove James wrong.

Five months after visiting Posner, I was wheezing my way up a long, stone flight of steps leading to a tall, ornate stupa perched on the side of a Colorado mountain. Against a clear blue sky, the 108-foot Buddhist reliquary was blindingly snow white, splashed with intricate carvings painted in a rainbow of hues. Far below, spread across a forest clearing lay a scattering of low buildings, a Buddhist retreat that looked part summer camp, part conference center, hidden in remote country one hundred and twenty miles northwest of Denver. Having only acclimatized for a day, I felt queasy and scattered as I reached the top. But I removed my shoes, tiptoed into the stupa, settled onto a cushion on the floor among a dozen statue-still people sitting lotus-style before an immense gold Buddha—and vainly tried to meditate for the first time in my life. In-out-in-out-*relax*-and-focus-on-each-and-every-breath. What-does-this-have-to-do-with-attention?

This other-worldly spot, far removed from the barren confines of most psych labs, was nonetheless the site of history's most ambitious scientific study to date of attention training. The collaborative brain-child of an intense Buddhist contemplative scholar and a feisty New York-born neuroscientist, the study was so intricate and original that its results just might wind up someday helping redefine our under-standing of attention. The main aim was simple—to study how three months of cloistered meditation practice affected the attention skills and overall social and emotional health of a diverse group of thirty people. The execution, however, was complex: six young neuroscien-tists, led by Clifford Saron of the University of California at Davis, worked fourteen hours a day testing the contemplatives on a plethora of measures ranging from heart rate, stress hormones in saliva, and indicators of immune function in blood to Stroop and other attention tests.[27] Three times during their stay, the meditators endured nine to ten hours of testing in twin specially built laboratories tucked in the basement of their dorm-style home. ("Some of them really hate it," confided a graduate student.) In addition, they kept daily diaries that were to be combed later for clues to their moods and insights. Mean-while, a randomly selected, thirty-person control group, who com-prised the basis of a second experimental retreat in the fall, flew in regularly for the same tests. By the time I arrived in May at the Shambhala Mountain Center for the initial retreat's closing days, Saron and his crew had compiled seventeen hundred hours of data that will take years for his research team and two dozen scientific collabo-rators to sift through. Dave Meyer, the expert on multitasking, had just arrived for a tour, curious to see the daredevil, high-altitude effort. Saron was sleep-deprived, his young researchers were exhausted from overwork, and the meditators were on edge in expectation of soon trading a television-, computer- and phone-less quiet for a noisy, wired normality. Even talking was novel; the retreat had been almost entirely silent. But if a secret to sparking a society that values attention is realizing that we can strengthen this set of skills, then this highly unusual effort was not so out-of-the-way. The Shamatha Project, named after a Buddhist term for cultivating attention, sits squarely at the crossroads of a set of discoveries that could make attention training, meditative or not, a normal part of life. In short, we're on the verge of learning how to meet head-on a distraction-charged world.

Consider first the Cinderella story of neuroplasticity. William James surmised that habits of mind and body shape our nervous systems, "just as a sheet of paper or a coat, once creased or folded, tends to fall forever afterward into the same identical folds."[28] Much later, scientists such as Eric Kandel painstakingly began to uncover how learning and memory boosts and changes synaptic functioning. But only recently have we begun to fully discover the exciting and scary story of the brain's extraordinary plasticity. For generations, the adult mammal brain was written off as basically fixed: neuroscientists assumed that no new neurons could grow and that each chunk of a healthy brain has one role to play in life, explains science writer Sharon Begley. These assumptions have meant that until recently stroke patients commonly underwent loads of physical therapy, but their broad cognitive deficits often were ignored.[29] Similarly, no one would have thought to try and rewire the brain of a dyslexic or a depressive. "But the dogma is wrong," relates Begley. "The adult brain, in short, retains much of the plasticity of the developing brain, including the power to repair damaged regions, to grow new neurons, to rezone regions that performed one task and have them assume a new task."[30] How our habits do shape us. Like fairy godmothers pulling rag-girls from the ashes, the deaf can recruit their dormant auditory cortex for *vision* tasks, while the blind can process language, including reading Braille, through their visual cortex.[31] Exercise seems to boost the efficiency of the anterior cingulate by increasing blood flow and the growth of new neurons.[32] Posner's finding that parenting can influence gene expression also underscores how environment shapes cognition. "Who we are and how we work comes from our perceptions and experiences," neuroscientist Helen Neville told the Dalai Lama in 2004 at a scientific meeting that he convened on neuroplasticity.[33] Of course, Neville was preaching to the choir. Buddhism has always held "consciousness as highly amenable to change," wrote the Tibetan leader in *The Universe in a Single Atom*.[34] In fact, shaping minds is the raison d'être of this spiritual tradition. This is why I'm sitting up in the cloud line of Red Feather Lakes, Colorado, stupefied but eager to learn how to quiet-my-racing-mind-let-go-of-distractions-and-return-to-my-breath.

Buddhism is empirical, scientific, psychological, and devoid of supernatural belief, writes religious scholar Huston Smith. "Instead of beginning with the universe and closing in on man's place in it, Buddha invariably began with man, his problems, his nature, and the dynamics of his development," writes Smith in *The Religions of Man*.[35] To Buddhists, the astonishing potential of the mind is the source of man's joy and troubles, and intense mental training—meditation—is the path to bettering oneself and the world.[36] Meditation isn't about emptying the mind, as is commonly thought, but instead involves concentrated efforts to explore and control its workings. It's a kind of twenty-five-hundred-year-old mental gymnastics, performed not as an end in itself, but rather as a stepping stone to becoming more compassionate, calm, and joyful—traits few would argue are overly abundant in our era. By itself, mindful breathing is a "completely useless skill," observes B. Alan Wallace, a Buddhist scholar.[37] Yet this beginner's technique, one of a huge roster of meditation practices, opens the door to controlling and then training the mind, just as musicians practice scales and athletes do drills to build up to a superior performance. Like James, we tend to think that we're born focused or scattered. We try not to daydream or zone out, obsess, or explode too much, but hardly spend time, even in this age of revolutionary brain-science, pondering techniques for managing our thoughts. Our brains seem like little machines tucked away in our skulls, intricate computers fueled by complex software, out of reach to reprogram or even understand. That, of course, is the twentieth-century legacy of Cartesian dualism, as well as the sobering implication borne from our increasing realization of the power of the subconscious brain. In short, we *aren't* always in charge. But that doesn't mean that we have to vacate the driver's seat.

The moon was full, and distant thunder rumbled across the mountains as Alan Wallace took a seat at the head of the hushed Hall of Sacred Studies. Several dozen staff from the Mountain Center sat on meditation cushions on a bamboo floor to hear him and Saron break their silence for a farewell talk about the Shamatha Project. Wallace is a wiry, incisive California-born, eclectic scholar and former Buddhist monk who has capped a globe-trotting childhood with a lifetime of study and teaching in Tibetan spiritual realms, partly under the guidance of the Dalai Lama. He dispenses teachings like a verbal water cannon in a kind of shorthand perhaps inspired by his multilingual beginnings. "Enthusiastic I was, but so uptight!" he writes in *The Attention Revolution: Unlocking the Power of the Focused Mind*, recalling an early solo retreat in a rat- and bug-infested Indian mountain hut long ago.[38] Known as an astute synthesizer of a vast, complex tradition, Wallace juggles intellectual quests from unlocking consciousness to founding a science of contemplation. Yet a foremost devotion is teaching shamatha—the cultivation of attention, for attention has been the lynchpin of exploring the mind through meditation for millennia. Without attentional skills, the deeper recesses of the mind stay hidden and intellectual control remains a pipedream. This is a central tenet of Tibetan Buddhism, yet Wallace is frustrated that some practitioners skip over the hard work of honing attention, opting for an emphasis on just being "in the moment." Wallace is a purist, drawing on traditional texts to define mindfulness as an awareness that includes a steady, unwavering focus. He also speaks of honing "presence," or a kind of engaged awareness of your surroundings and "metacognition," an overarching ability to watch and understand your own mind. The words differ, but the similarities are striking: for twenty-five hundred years, this multifaceted meditative tradition has been refining networks of attention that neuroscience has just begun to plumb. Like Posner, Wallace offers us a shared language of attention to quiet the Babeling tongues of distraction.

Can attention be trained? To Wallace, who has mediated ten thousand hours in solitary retreats over thirty-five years, the answer is already in. But he believes that science can help unpack the intricacies of meditative attention, just as meditation is the perfect canvas for scientifically understanding how attention can be shaped. At the meeting, he and Saron emphasized that they aren't studying Buddhism but rather the attentional training within this tradition, along with the cultivation of compassion and other Buddhist "qualities of the heart" as objects of focus. "It's fundamentally a non-denominational pursuit," quipped Saron. Added Wallace, "The point is I think there really can be a two-way traffic here." They bantered about their differences: science asserts that mind emerges from the brain, and Buddhism holds that thought is much more than just chemistry and biology. "The fact that we do have very different world views has made it fun," said Wallace, a long, burgundy meditation shawl draped across his shoulders. Later, I talked to a half-dozen project participants, a diverse, mostly middle-aged group who came from around the world, giving up jobs or taking leaves and paying $6,000 each to be tested, taught, and closeted for three months. Naturally, most report feeling more focused and calm. After all, they'd had a rare chance to step away from their frenetic lives to unhurriedly heal and reflect. "I couldn't help but think, 'Oh God, just to focus on one thing!' You know, what a relief that would be," said Cass McLaughlin, a can-do, dusty-blonde outreach coordinator for the University of Minnesota's Center for Spirituality & Healing, as she recalled one of her motivations for joining. "It took six weeks just to unwind." At the retreat, she learned "to get to a calmer, more still place than I've ever been before."[39] Is this perhaps the "vacation effect," destined to evaporate as soon as she returns? Wallace soberly ended the meeting with these and other tough questions. "What's the generalizability or transferability of those attention skills that have been cultivated? And if it can be trained, is there a very low ceiling, a high ceiling?" he intoned, as some among the believers fidgeted uneasily. "These are really fundamental questions for which scientifically right now we don't have very clear answers." And yet, the answers are trickling in.

Four days later, I met Saron and Meyer again, this time at a bustling old monastery-turned-conference center set on a wooded cliff overlooking New York's Hudson River. Dozens of neuroscientists were gathered for an annual five-day meeting organized by the Mind and Life Institute, a Colorado nonprofit that supports the scientific study of meditation. Meyer was one of the meeting's masters of ceremonies; he was drawn to meditation after his son's death, although he's kept his research separate. Also presiding was Richard J. Davidson, known for his studies of how expert meditators regulate emotion. Just two decades ago, no career-minded scientist would have pursued this work. Now, gleanings from contemplative brains make routine headlines. Hospitals offer mindfulness-based stress reduction programs that were pioneered by molecular biologist Jon Kabat-Zinn to treat ills from chronic pain to skin diseases. Neuroscientists are studying nascent efforts to bring meditation into classrooms. Attention, however, is a new research focus in this budding subfield of brain science, and so the main meeting hall was packed as the neuroscientist Amishi Jha took the stage to relate her landmark findings.[40]

Jha compared seventeen novices taking an eight-week introductory course in meditation with seventeen experienced meditators studying at a one-month retreat. Before and after the courses, she gave them Posner's Attention Network Test, a simple twenty-minute computer exam that measures a person's efficiency in each of the three networks. The veterans demonstrated better executive attention than newcomers at the start, then showed greater improvement in the alerting network after their retreat. The novices, in turn, made robust gains in orienting, suggesting to Jha that meditation may sharpen focus first and then a wider wakeful awareness. But more than the training effects, what surprised Jha was the rare evidence of carryover: meditation boosted people's prowess on laboratory tasks that were wholly distinct from the specific practice of mindful breathing. "It would be as if learning to ride a bike helped you be a better tightrope walker," Jha, an assistant professor at the University of Pennsylvania, told me later. "There's something about keeping steady that is related, but mostly they are quite different activities." In her study, just eight weeks of meditation training boosted scores on tests of a type of orienting that involves spatial skills. Somehow, pulling one's attention back to the breath again and again helped hone the flashlight of attention. "The whole point is generalizability in training," breathlessly related Jha, who found in meditation a few years ago the cognitive booster rocket she desperately needed to cope with life's stresses. "If you spend thirty minutes a day and it makes a difference in the quality of your attention, that is powerful."

We are not born with a fixed allotment of focus, William James. Mounting evidence of our attentional malleability is inspiring Posner and Rothbart to urge schools to train children in attention—before it's too late. If adult focus can be sharpened in just eight weeks, why not tackle young networks at their most sensitive stages of development? A few years ago, Mary Rothbart returned from a conference where she'd had an "a-ha" moment while hearing Georgia State primatologist Duane Rumbaugh speak about his astonishing success training monkeys on computer tasks. No one thought it could be done, but, as Rumbaugh and collaborator David Washburn have drily noted, "[F]ortunately, the rhesus monkeys had not read the literature."[41] Joystick in hand, the monkeys went to town, hammering away at Stroop-like and other exercises carefully designed to test their capacity to learn. To the researcher's surprise, the tiny players also seemed to grow more focused and cooperative, qualities that captive-born monkeys notoriously lack. This unexpected result inspired a new research focus for Washburn and opened up a remarkable new line of investigation for the Oregon team. Posner and Rothbart turned the monkey business into five kid-friendly exercises, including the one that six-year-old Emma had demonstrated for me in Posner's lab. The activities tone executive attention skills including working memory, self-control, planning, and observation. Six-year-olds trained in seven half-hour sessions showed a pattern of brain wave activity in the anterior cingulate similar to that of adults and a marked gain in executive attention. Four-year-olds jumped six points on IQ tests, driven by sharp upswings in culture-free aspects of intelligence, such as fluid thinking and nonverbal reasoning. Among both age groups, those worst off in attentional skills at the outset gained the most from the training.[42] The implications are vivid. First, the results show how some carefully designed computer tasks can be vehicles for powerful learning. We can't expect computers—and especially games—automatically to be our teachers, regardless of the content, yet they can be a *means* of learning, especially in the hands of a sage thinker, such as Posner. More importantly, Posner's results show that we are beginning to learn how attention can be both *understood* and *trained*. "The fact that we can now trace the development of neural networks in the human brain that are at the heart of the educational experience means that we can try to change them for the better," Posner reflected in a presidential-invited address to the American Psychological Association's annual meeting. "We should think of this work not just as remediation but as a normal part of education."[43]

Can attention be trained even in someone whose attentional networks are off-kilter? Once considered biologically fixed and treatable only through symptom relief, ADHD is increasingly seen as a disorder sparked by neural network weaknesses that can be strengthened.[44] This means that for the first time, we can attack attention deficits head-on, *by boosting attention*. For example, five weeks of daily half-hour sessions on a computer program called RoboMemo, developed by Swedish neuroscientist Torkel Klingberg, bolstered working memory skills in kids with ADHD. Klingberg's hypothesis is clever: strengthen verbal and visuospatial working memory, our cognitive cupboards, and you'll likely find a back door to toning executive attention. If you can't recollect the details of the moment, you can't carve a better future from the rough gem of the present. And RoboMemo seems to inspire carryover: the program offers some improvements in reasoning, conflict resolution, and other executive functions. (One small problem: in pilot studies, Swedish kids worked hard for stickers as a reward for doing the training, but US kids found them ho-hum.)[45]

In another corner of the world, Leanne Tamm adapted Posner's training for the executive network in Los Angeles-area preschoolers at risk for ADHD, with impressive early results. But her new work in Dallas is most intriguing. Setting aside computers, she's coaxing eight- to twelve-year-olds with ADHD through the kind of attention-training tasks developed in the 1980s for people with brain injuries, such as hitting a buzzer when they hear a certain word on a storytape played in a noisy room. And she is again working with three- to seven-year-olds, this time schooling families in attention's varied forms and teaching them to reconnect over block stacking, card matching, or word games that bolster attention, self-regulation, and memory. "We're trying to build back the family relationship, rather than say, 'here's a CD, put it in the TV,'" says Tamm, an assistant professor of psychiatry at the Southwestern Medical Center of the University of Texas. "It's a society where there's a limited time to build in playdates with your kids."[46] Not far into her multiyear study, teachers and parents report exciting changes in just eight weeks, and the children are thrilled both to receive the attention they crave and to begin to understand the nebulous concept. "Kids are always told to pay attention, but they don't know what that means," says Tamm. "One of the most critical elements is giving them a common language for what it means to pay attention." A *language of attention*. Only when we speak this language can we bestow on others the irreplaceable gift of our attention.

"Attending to, caring for, watching over, looking after—there is a real sweetness to that," said Alan Wallace. Clad in a sport shirt and chinos, Wallace sat cross-legged on a chair in his room at the retreat, mulling over vocabularies of attention with me, savoring each as if tasting a fine old wine. Before I headed home, he proffered a last thought. "If a person leaps in and sacrifices his life—you leap in and save a baby and then you die—you've given your whole life in one piece," said Wallace in characteristic staccato-speak. "That's a wonderful sacrifice. Greater love hath no man than he who lays down his life for another. Jesus, right?[47] So that's pretty good. But when we give another person our attention, we've giving away that portion of our life. We don't get it back. We're giving our attention to what seems worthy of our life from moment to moment. Attention, the cultivation of attention, is absolutely core. It really is the lynchpin. It is the key."

On the way back to Denver to catch my flight, I got lost in the mountain foothills and began to panic, seeing few houses or cars for miles. At last, a man in a pick-up truck rumbled toward me, and I flagged him down. "Follow me," he said with a warm smile when I asked for help. "I'm going that way."

Wallace's parting words pose a conundrum that haunts any student of attention: attention to what? Certainly, the question of attention's forms is often entangled with issues of content. In other words, how we pay attention can subtly shape what we pay attention to. Splitting one's focus between a work project and one's child demotes both to half a priority each. Detachment dulls the colors of the world to a blurred gray. With eroding attention, we settle for the quick fix, surface observation, black-and-white thought. We get what we pay for, in a sense. Or, considered from another angle, the question becomes— what is a distraction? The *Oxford English Dictionary* defines distraction as a "drawing away (of the mind or thoughts) from one point or course to another; diversion of the mind or attention, usually in adverse sense."[48] But what constitutes a distraction? Are we paying attention to the screen or are we distracted by it? The answer is slippery, tantalizing, daunting, for distraction is in the eye of the beholder.

Consider pain. Pain *commands* our attention to ensure our survival. The perception of pain involves a network of at least five areas of the brain, including the ventral half of the anterior cingulate.[49] The involvement of executive attention's core player ensures that pain is both difficult to ignore and often a product of our perception. Think about pain and it worsens. Believe that you've found relief—pop a placebo—and you've switched on the brain's endorphin-based pain modulation networks. Distraction, as well, is known in medical circles to reduce pain. The spotlight of attention swivels away from the source of pain, in a sense relegating it to background noise. University of Washington psychology researcher Hunter Hoffman has taken advantage of the potency of distraction by developing virtual reality games for burn patients, as well as for people with phobias and post-traumatic stress disorders. In "SnowWorld," burn victims who are undergoing horrifically painful daily wound care can float through a mountainous arctic wilderness, lobbing snowballs at snowmen and penguins who burst or dissolve into ice-blue shards, all to the soothing tune of "Graceland" and other Paul Simon songs. Playing the game lessens pain dramatically, he's found. "The cardinal virtue of virtual reality—the ability to give users the sense that they are 'somewhere else' can be of great value in a medical setting," says Hoffman, an affable man with a dreamy air whose games are available in a growing number of hospitals.[50] When we *choose*, distraction can be a welcome abductor of our attention, as Mischel's adept young marshmallow shunners illustrate. But as William James observed in *The Principles of Psychology*, "The art of being wise is the art of knowing what to overlook."[51] Wisdom is not fed upon a diet of distraction.

Peer further into the story of distraction, and you'll find the word hails from the Latin for "pull asunder," an etymology that inspired an unfortunately now obsolete sense of the word. From the sixteenth to the early nineteenth centuries, distraction alternatively meant a rending into parts, a scattering, a dispersion. "While he was yet in Rome, His power went out in such distractions, As beguiled all Spies," says a soldier of Octavius Caesar in Shakespeare's *Anthony and Cleopatra*, written in 1606.[52] The *OED*'s lexical sleuths trace the first use of our familiar definition of the word, in turn, to the pages of a slim devotional treatise, *The Myroure of oure Ladye*, likely written between 1460 and 1500 by Thomas Gascoigne, a theologian and Oxford chancellor. No one can be certain of the date because the text is so rare that only fragments remain; the book is distracted, you might say. Compiled for a wealthy convent that was established Thames-side at Islesworth near London in 1415, the *Myroure* is a handbook of reli-

gious life for the Sisters of Sion. The book offered an English translation of their nine daily Latin chapel services and instruction on other
details of their highly circumscribed lives. Except during services, the
nuns were silent; hand gestures were to be used frugally. Strict hierar-
chies of authority prevailed. And "distraccyon" or "thynkeynge of
other thynges" during worship naturally was roundly discouraged.[53]

Still, the *Myroure* endeavors not to flatly forbid daydreaming but to
teach the art of attention. Its passages on distraction are warm, meditative, and unexpectedly instructive even to a reader far removed from a
wimple-clad life of prayer. Paying attention takes effort, Gascoigne
cautions, and becomes a habit only through long, patient cultivation of
the mind. Attention during worship is multilayered, involving a focus
on the words of the service, their literal and inward meanings, and the
ongoing ritual. Sometimes distraction is the work of the devil or both-
ersome fellow worshipers who are like brambles in the corn, stifling
others' good growth, the author asserts. But our mind also wanders
because we're preoccupied with "bodily or worldely or vayne
thynges," or because we lack the will—the "kepyng of the harte"—to
pull our focus back to our first priority. No penance is needed if,
through "neglygence or fraylte," your thoughts occasionally stray
during church, concludes Gascoigne. "He is not bounde to say that
seruice ageyne." More than five hundred years after the *Myroure*'s sage
teachings, psychologists are discovering mind wandering as a rich and
overlooked area of study. Also called "zoning out" or "mind popping,"
mind wandering is a turning inward, a decoupling of our attention away
from the outside world, that most commonly occurs when the task at
hand is easy, lengthy, and familiar, studies by Jonathan Smallwood and
Jonathan Schooler show.[54] It happens from 15 to 50 percent of the time
and invariably takes us by surprise, yet it may, as many consummate
daydreamers know, stoke problem solving, they argue. Our ever-
humming mind may hijack our focus in order to break into our mental
airwaves with an urgent bulletin about a different goal than the one in
front of us. Here again, we see that attention never will be completely
our docile, cognitive dog, rigidly obeying our wishes. But perhaps the

pragmatic sisters knew this well. They had their "spiritual amuse-ments," including readings for the "edyfyeng of [their] sowles" during meals, writes a nineteenth-century scholar who compiled one edition of the book. Attention was central to life in the medieval order, but dis-traction had a rightful place on the sidelines.

Attention is not always within our control. The unexpected, the changeable, the novel, even the habitual in life abduct our focus, intrude upon our awareness, and pull us off course for a time. Atten-tion is like a second skin, a meeting ground for our ever-present grap-pling with external and internal worlds. Yet used well and nurtured carefully, our networks of attention are our foremost means to shaping our lives. These networks give us extraordinary ways to master our-selves and our environment, offering the key to growth, connection, happiness. Accepting a culture of eroding attention relinquishes this potential for sculpting our individual and collective futures. To para-phrase Walter Mischel's caveat on will, we don't always want to exer-cise our highest powers of attention, yet if we cannot focus, observe, or judge well, the choice is lost. Will we slip into a dark age of dis-traction? At journey's end, I searched for final clues in contrasting realms of art and science where attention is nonetheless similarly dissected, rekindled, and venerated. To reverse a darkening time, we must understand, strengthen, and lastly *value* attention.

Step into Jacob Collins's dark, cavernous studio and art school in a converted Manhattan carriage house and you enter a seemingly bygone world.[55] The gray-brown walls are hung with ascending rows of paint-ings. Gleaming white plaster casts crowd a flight of shelves. Students stand or perch on stools before a jumble of easels clustered around a seated nude on a floodlit dais. In the dim light, the room is reminiscent of the kind of nineteenth-century atelier that I've seen only in paintings. Then, out bursts Collins from his studio in the back of the building, and the scene is startlingly animate. A generous yet somewhat combustible forty-three-year-old with sandy, tousled hair and paint-splattered clothes, Collins burns with visions of resuscitating an art world long past. Great-nephew of the critic Meyer Schapiro and scion of a line of New York artists and thinkers, he is a counter-revolutionary, leading a movement to rekindle the tradition of realism in an era still dominated by abstraction. And in just a few decades, the effort is gaining. Dozens of similar ateliers have sprung up in the United States and Europe. Collins, whose luminescent works sell for upwards of $300,000, recently helped found two other art schools to teach still-life, portraiture, and landscape largely through years-long, apprentice-style programs. "The story of art just went up for a re-edit," writes critic James Panero.[56] Perhaps. And yet, whether or not you support his artistic crusade or prefer contemporary art to traditional realism, Collins's efforts offer us the hope of a shared revisioning—of our culture of distraction. I am on his doorstep not to enter an art world fray but to learn from a master of focus, an artist who brings the study of attention to life.

Collins knew as a boy that he wanted to paint, yet he found learning the craft frustrating. He'd always been fascinated by "things that were magically well done," such as calligraphy, architectural carvings, drawings, master paintings. But late twentieth-century art schools favored rapid sketching, creativity in bursts, and impression-istic techniques that Collins felt emphasized speed over patience. Nimble, even distractible ways of focus have a place in life, empha-sizes Collins, as long as we can still cultivate their complements: forms of deeper, sustained, whole attention. After graduating with a history degree from Columbia in 1986, Collins studied art in New York and France, then began to tap into a tradition of largely forgotten techniques being revived by artists such as Anthony Ryder, who spends up to seventy hours on a single drawing. "I was dazzled," recalled Collins, who first met Ryder, now a close friend, when they studied together. "The longest I had spent on a drawing was three to four hours. I had been taught that everything was made in a fit of cre-ativity." Collins remains unrepentantly out of step with the main-stream art world, and his place in its history is uncertain. Yet like Alan Wallace or Mike Posner, he gives us vocabularies of attention that can help counterbalance our easy fluency in the language of distraction.

Now Collins teaches his students not just how to paint but how to step away from an "attention-free culture, with channel-flipping, the manic pace, the sort of auto culture causing you to jump here, run there." He introduces his students to techniques of attention, from organizational frameworks of planning a work, to scanning a scene two-dimensionally to see only shapes, or projecting oneself into the imaginary three-dimensional space behind a drawing to experience form, light, and tone. Picking up a notebook and roughly sketching a human thigh, he talked of "modules of knowledge" such as anatomy and volume, of finding the arabesque in nature, of losing yourself, flow-like, in the process of creation. "They're all ideas that require a tremendous amount of attention," he said. "I find myself balancing back and forth between that narrow, infinitely particular aspiration (which I love), and this potentially overly distractible breadth of thinking. You want to make it how it is, not just how it looks." He noted that the artists he emulates—from Frederic Church to Paul Delaroche—came from a world with time for reflection and close observation of nature, so the magic of their art stems from their tech-niques and abilities as well as the "ambient presence" of attention in their world. Then he winced. His cell phone was ringing for the third time in an hour—once ignored, twice taken. "My life is an infinity of distractions," he apologized, talking briefly with a caller before recov-ering the thread of our conversation. An essential part of his life work has been schooling himself in patience and focus, said Collins, who lives upstairs from the school with his wife, a novelist, and their three children. "I have a jump-around quality, and partially because of that, I've cultivated the very opposite. I've worked hard to be centered, and I've wondered whether that's almost like a personal obligation that everyone has." Creating a culture of attention in a distractible world may be Jacob Collins's most priceless artwork yet.

Blink. Clang. Click. Crash. I have moved from the quiet sanctum of the atelier to the chilly confines of a cacophonous brain-scanner. I am flat on my back in a tubular machine, head encaged in a kind of catcher's mask, hands strapped onto mitt-shaped mini-keyboards, eyes glued to a computer screen visible behind me via a tiny mirror. The MRI scanner growls, whirrs, bangs, even shakes, but I try to focus on an empty little box winking at me on the computer screen. *"Warning!" it seems to say. "The action begins here."* Today, I am back in the machine, subjecting myself this time not to the eerie symbolism of the Technology Plays but to a scientific readout of my attentional prowess. Here at the Mount Sinai School of Medicine, just a few miles from the studio of Jacob Collins, I am taking the Attention Network Test, a recently developed kind of half-hour psychological biopsy of my alerting, orienting, and executive networks, as part of an ambitious new study to discover how individuals pay attention well or poorly.[57] Having steeped myself for years in the mysteries of attention amid a building climate of distraction, I'm adding a little bit to the light—and gaining a peek at my own inscrutable mind. Will I flunk, and how can a test so simple shed light on a system so complex? Funny how those little coffinlike boxes look just like the eyeglasses of the rumpled young scientist who has shepherded me through hours of genetic, psychological, intelligence, and now attention testing. The final exam is just beginning, and I'm already daydreaming. The test "can be easily performed by children, patients and monkeys," write the inventors. This is not good.

Developed by neuroscientists Jin Fan and Bruce McCandliss under the guidance of Mike Posner, the ANT is a widely used metric whose simplicity belies its formidable explanatory powers. A test taker essentially faces only one challenge: report the direction of an arrow on-screen by pressing a button with her right or left finger. Easy, but for the many on-screen complications. Each "target" arrow is flanked by a pair of arrows. Sometimes, the target and its brethren all point in the same direction. Sometimes, like a mixed-up Radio City Rockette, the target faces the wrong direction—a situation that gives the test taker pause, just like a Stroop word trip-up. Such incongruity measures your executive attention, that is, your ability to keep your eye on the ball in the face of distractions. All the while, the empty boxes surrounding the arrows flash—at times quickly, other times slowly—to warn whether the school of arrows will appear on the left or right side of the screen. Those boxy little wake-up calls speed up your responses, except when both boxes flash at once or one betrays you by luring your gaze to the wrong side of the screen. Click by click, the multitudinous variations ingeniously test your ability to stay alert, to orient your focus, and to overlook distractions. At the same time, the scanner takes snapshots inside your pulsing skull, essentially measuring both brain structure and the relative concentrations of some compounds in your brain as you think. (Active brain areas are suffused with oxygenated blood, which has magnetic properties that can be tracked by the scanner's enormous magnets.) After 288 blasts of arrows and boxes, I have a pounding headache, a sinking feeling, and a neat trio of scores. What do they show?

We are on the cusp of an astonishing time, and on the edge of darkness. *I score an average reaction time of 77 milliseconds on alerting and 52 on orienting, figures calculated from how well I refocus or react when tipped to a change. I can react to a shifting world.* We now hold the potential to know, shape, and utilize a full quiver of attentional skills to combat a spreading culture of distraction. *My score of 140 milliseconds on executive attention is on a par with the norm. I can see the way forward, not perfectly.* We can create a culture of attention, recover the ability to pause, focus, connect, judge, and enter deeply into a relationship or an idea, or we can slip into numb days of easy diffusion and detachment. What did I learn from this test, this voyage, this quest to understand the mysteries of attention? We are on the edge. The journey is just beginning. Will we cultivate a renaissance of attention? The choice is ours.

ACKNOWLEDGMENTS

In writing this book, I asked many people for the gift of their attention. I am deeply grateful to all those who generously shared their time, stories, wisdom, and thoughts with me.

Among my most wise and patient teachers were Mike Posner, Dave Meyer, Bruce McCandliss, Norbert Elliot, Alan Wallace, and Elinor Ochs. I also benefited greatly from the insights of Amishi Jha, James McGaugh, Dan Anderson, Brian Wansink, Art Kramer, Kevin Guise, Inga Treitler, Frances Kuo, Bill Gehring, David Fencsik, Mike Liebhold, Jim Mulvaney, Abby Smith, and Rebecca Baron.

Others graciously supported this work in myriad ways. Many thanks to Mark Bartlett and the staff of the New York Society Library, to David Smith at the New York Public Library's main research division, and to Stuart Basefsky, a senior reference librarian at Cornell University. A number of people astutely commented on the manuscript, in part or in full: Wallis Miller, Karen Cornelius, Norbert Elliot, Peter Cavaluzzi, Jim Burns, Allison Cheston, John Fossella, and Ozlem Ayduk. (All errors are, of course, my own.) I also appreciated the kind help and encouragement of Karen Smul, Alexandra Kahn, Andrea Saveri, Ross Barrett, Ellen Galinsky, Carol Guensburg, Arlene

Johnson, Amy Klobuchar, Sarah Marvin, Michael Hagner, Sharon Posner, Rosemary Bowen, Jim and Ann Marie and all the Hitchcock clan, Clista Dow, Merrill Smith, Lisa Jackson, Mark Gallogly, Lisa Brainerd, and my book club. Special thanks to Cindy Wentworth, Ellen Feldman, and Lise Strickler for keeping me sane.

Emma Sweeney is a wise, attentive agent, and I am thrilled to work with her and her wonderful associate Eva Talmadge. I am very grateful for the support of Jonathan Kurtz, Steven L. Mitchell, and Jill Maxick and so many others at Prometheus. The book benefited tremendously from Linda Regan's sage editorial eye and her enthusiasm for my vision.

Time and again, the fascinating subject of attention captured my attention fully, yet my family cheered me on and soldiered on, with understanding and patience. John, Emma, and Anna shared the ups and down of the research and writing, endured my absences and worries, did more than their share of the chores, and still met me at the door with hugs and "welcome home" signs. Thank you.

ENDNOTES

Note: All URLs are current as of October 2007.

INTRODUCTION

1. Daniel Ho, "*1+1=0.*" Also Richard Dresser, "Greetings from the Home Office," and Malcolm Messersmith, "Chip" (presented at the State University of New York, Albany, November 2003).

2. Jay Palmer, "High-Wired Act," *Barron's*, April 17, 2007, p. 27. Also *Books in Print 2007–8* (New Providence, NJ: Bowker, 2007), 1:vii.

3. William James, *The Principles of Psychology*, ed. Frederick Burkhardt (Cambridge, MA: Harvard University Press, 1981; originally published, 1890), 1:381–2.

4. Barbara Tuchman, *A Distant Mirror: The Calamitous Fourteenth Century* (New York: Knopf, 1978), p. 55. Also Chiara Frugoni, *Books, Bank, Buttons and Other Inventions of the Middle Ages*, trans. William McCuaig (New York: Columbia University Press, 2001, 2003), pp. ix–x.

5. Thomas Cahill, *How the Irish Saved Civilization: The Untold Story of Ireland's Heroic Role from the Fall of Rome to the Rise of Medieval Europe* (New York: Nan A. Talese, Doubleday, 1995), p. 181.

6. Dan Stanislawski, "Dark Age Contributions to the Mediterranean

Way of Life," *Annals of the Association of American Geographers* 63, no. 4 (1973): 397–410, here, p. 398. Also Anthony Snodgrass, *The Dark Age of Greece: An Archeological Survey of the Eleventh to the Eighth Centuries BC* (Edinburgh: University of Edinburgh Press, 1971), pp. 2, 21, 363, 367, 381, 399–402.

7. Kenneth Clark, *Civilization: A Personal View* (London: British Broadcasting Corp. and John Murray, 1969), pp. 14, 17.

8. Umberto Eco, *Travels in Hyper Reality: Essays*, trans. William Weaver (San Diego: Harcourt Brace Jovanovich, 1986), p. 73. Jane Jacobs, *Dark Age Ahead* (New York: Random House, 2004), p. 3. Harold Bloom, "Great Dane," *Wall Street Journal*, April 20, 2005, p. A16.

9. Hallowell quoted in Kris Maher, "The Jungle: Focus on Recruitment, Pay and Getting Ahead," *Wall Street Journal*, March 2, 2004, p. B8.

10. Ellen Galinsky et al., *Overwork in America: When the Way We Work Becomes Too Much* (New York: Families and Work Institute, 2005), pp. 2–4.

11. Gloria Mark, Victor Gonzalez, and Justin Harris, "No Task Left Behind? Examining the Nature of Fragmented Work," *Proceedings of the Conference on Human Factors in Computer Systems* (2005): 321–30. Also Victor Gonzalez and Gloria Mark, "Constant, Constant Multi-tasking Craziness: Managing Multiple Working Spheres," *Proceedings of Conference on Human Factors in Computing Systems* (2004): 113–20.

12. Victoria Rideout and Donald Roberts, *Generation M: Media in the Lives of Eight to 18-Year-Olds* (Menlo Park, CA: Henry J. Kaiser Family Foundation, March 2005), pp. 6, 23.

13. Deborah Fallows, *Search Engine Users* (Washington, DC: Pew Internet and American Life Project), January 23, 2005, pp. iii–iv, 27, http://www.pewinternet.org/pdfs/PIP_searchengine_users.pdf. Also Patti Caravello et al., *Information Competence at UCLA: Report of a Survey Project* (Los Angeles: UCLA Library Instructional Services Advisory Committee, Spring 2001), pp. 1–3, http://www.library.ucla.edu/infocompetence. Hoa Loranger and Jakob Nielsen, *Teenagers on the Web: Usability Guidelines for Creating Compelling Websites for Teens* (Fremont, CA: Nielsen Norman Group, 2005), pp. 5–6, 12–13, 17–18.

14. *Problem Solving for Tomorrow's World* (Paris: Organization of Economic Co-operation and Development, 2004), pp. 40–42, 47, 144, http://www.oecd.org/dataoecd/25/12/34009000.pdf.

15. H. Persky, M. Daane, and Y. Jin, *The Nation's Report Card: Writing*

2002 (Washington, DC: US Department of Education, Institute of Educational Sciences, 2003), pp. 11, 19, 21, http://nces.ed.gov/nationsreportcard/ pdf/main2002/2003529.pdf. Also Steven Ingels et al., *A Profile of the American High School Sophomore in 2002: Initial Results from the Base Year of the Education Longitudinal Study of 2002* (Washington, DC: US Department of Education, National Center for Education Statistics, 2005), pp. 25–26, http://nces.ed.gov/pubs2005/2005338.pdf. Also *Crisis at the Core: Preparing All Students for College and Work* (Iowa City: ACT, 2005), pp. 3, 24, http://www.act.org/path/policy/pdf/crisis_report.pdf.

16. Steven Johnson, *Everything Bad Is Good for You: How Today's Popular Culture Is Actually Making Us Smarter* (New York: Riverhead Books, 2005), pp. 41–46, 61–62.

17. Kaveri Subrahmanyam et al., "The Impact of Home Computer Use on Children's Activities and Development," *Future of Children* 10, no. 2 (2000): 123–44, here, p. 129. Also Sandra Calvert, "Cognitive Effects of Videogames," in *Handbook of Computer Game Studies*, ed. J. Raessens and J. Gudstein (Cambridge, MA: MIT Press, 2005), pp. 125–31. Also June Lee and Aletha Huston, "Educational Televisual Media Effects," in *Faces of Televisual Media*, ed. Edward Palmer and Brian Young (Mahwah, NJ: L. Erlbaum Associates, 2003), pp. 83–105. Also C. Shawn Green and Daphne Bavelier, "Action Video Game Modifies Visual Selective Attention," *Nature* 423, no. 29 (2003): 534–37.

18. Johnson also argues that nonverbal IQ scores are rising globally because more than ever we live in a world dominated by images, although others attribute the gains to better nutrition or even simply to the increasing complexity of society. See Ulric Neisser, *The Rising Curve: Long-Term Gains in IQ and Related Measures* (Washington, DC: American Psychological Association, 1998).

19. Interviews with Brendan, his classmates, and the librarian, June 2005.

20. Ian Parker, "Absolute PowerPoint: Can a Software Package Edit Our Thoughts?" *New Yorker*, May 28, 2001, pp. 76–87.

21. Edward Tufte, *The Cognitive Style of PowerPoint* (Cheshire, CT: Graphics Press, 2003). Clive Thompson, "PowerPoint Makes You Dumb," *New York Times*, December 14, 2003, p. 88.

22. Sherry Turkle, "From Powerful Ideas to Powerpoint," *Convergence* 9, no. 2 (2003): 24.

23. David Byrne, *Envisioning Emotional Epistemological Information* (Gottingen, Germany: Steidel Verlag, 2003).

24. Miller McPherson, Matthew Brashears, and Lynn Smith-Lovin, "Social Isolation in America: Changes in Core Discussion Networks over Two Decades," *American Sociological Review* 71 (2006): 353–75. Jerome Groopman, *How Doctors Think* (New York: Houghton Mifflin, 2007), p. 17. Victoria Rideout, Elizabeth Vandewater, and Ellen Wartella, *Zero to Six: Electronic Media in the Lives of Infants, Toddlers and Preschoolers* (Menlo Park, CA: Henry J. Kaiser Family Foundation, 2003), p. 4. Dimitri Christakis, Frederick Zimmerman, David DiGiuseppe, and Carolyn McCarty, "Early Television Exposure and Subsequent Attentional Problems in Children," *Pediatrics* 113, no. 4 (2004): 708–13. *Final Report: National Geographic-Roper Public Affairs 2006 Geographic Literacy Study* (Washington, DC: National Geographic Education Foundation, May 2006), pp. 6, 28, http://www.nationalgeographic.com/roper2006/.

25. Michael Posner and Jin Fan, "Attention as an Organ System," in *Topics in Integrative Neuroscience: From Cells to Cognition*, ed. James Pomerantz (New York: Cambridge University Press, 2007).

26. Michael Posner and Mary K. Rothbart, "Attention, Self-Regulation and Consciousness," *Philosophical Transactions: Biological Sciences* 353, no. 1377 (1998): 1916.

27. Mihaly Csikzsentmihalyi, *Flow: The Psychology of Optimal Experience* (New York: Harper Perennial, 1991), pp. 84–85.

28. Ibid.

29. Duane Rumbaugh and David Washburn, "Attention and Memory in Relation to Learning," in *Attention, Memory and Executive Function* (Baltimore, MD: Paul H. Brookes Publishing, 1995), pp. 199–219. Also e-mail communications with Washburn, September 2007.

30. Mary K. Rothbart and M. Rosario Rueda, "The Development of Effortful Control," in *Developing Individuality in the Human Brain: A Tribute to Michael I. Posner*, ed. Ulrich Mayr, Edward Awh, and Steven Keele (Washington, DC: American Psychological Association), p. 170.

31. *Oxford English Dictionary*, vol. 1 (Oxford: Clarendon Press, 1989), p. 765.

32. Cahill, *Irish Saved Civilization*, p. 59.

33. Lewis Lapham, *Waiting for the Barbarians* (London: Verso, 1997), p. 210.

CHAPTER ONE

1. Ella Cheever Thayer, *Wired Love: A Romance of Dots and Dashes* (New York: W. J. Johnston, 1880), p. 44.

2. Blaise Cendrars, *Transsibérien et de la petite Jehanne de France* (Paris: Editions des hommes nouveaux, 1913). Translated passages from Blaise Cendrars, *Prose of the Trans-Siberian & of the Little Jeannie de France*, trans. Tony Baker (Nether Edge, Sheffield: West House Books, 2001).

3. Tom Standage, *The Victorian Internet: The Remarkable Story of the Telegraph and the Nineteenth Century's Online Pioneers* (London: Walker and Co., 1998), p. 2.

4. James Katz and Mark Aakhus, eds., *Perpetual Contact: Mobile Communication, Private Talk, Public Performance* (New York: Cambridge University Press, 2001), p. 2.

5. James Gleick, *Faster: The Acceleration of Just about Everything* (New York: Pantheon Books, 1999), pp. 99–100.

6. Ibid., p. 90.

7. Standage, *Victorian Internet*, pp. 25, 102.

8. Karl Lamprecht, *Deutsche Geschichte de jüngsten Vergangenheit und Gegenwart* (Berlin, 1912), 1:172. Reprinted in Stephen Kern, *The Culture of Time and Space 1880–1918* (Cambridge, MA: Harvard University Press, 1983), p. 230.

9. Fernand Léger, "The Origins of Painting and Its Representational Value" (1913). Reprinted in Edward Fry, ed., *Cubism* (London: Thames and Hudson, 1966), pp. 121–27.

10. Kern, *Culture of Time and Space*, p. 77.

11. Paul Collins, "Love on a Wire," *New Scientist*, December 21, 2002, p. 40. Also Thayer, *Wired Love*, p. 25.

12. Standage, *Victorian Internet*, p. 133.

13. "Romances of the Telegraph," *Western Electrician* 9, no. 10 (1891): 130–31.

14. Kern, *Culture of Time and Space*, p. 314.

15. Teen quoted in Bonka Boneva et al., "Teenage Communication in the Instant Messaging Era," in *Computers, Phones and the Internet: Domesticating Information Technology*, ed. Robert Kraut, Malcolm Bryin, and Sara Kiesler (Oxford: Oxford University Press, 2006), pp. 201–18, here, p. 212.

16. Thomas Edison, *The Diary and Sundry Observations of Thomas Alva Edison*, ed. Dagobert Runes (New York: Philosophical Library, 1948), p. 233.

17. Pamela Thurschwell, *Literature, Technology and Magical Thinking: 1880–1920* (Cambridge: Cambridge University Press, 2001), pp. 22–23.

18. Ibid., p. 3.

19. Sarah Waters, "Ghosting the Interface: Cyberspace and Spiritualism," *Science as Culture* 6, no. 3 (1997): 414–43, here, p. 428.

20. Charles Herold, "What Fools These Avatars Be," *New York Times*, September 16, 2004, p. G1.

21. Jonathan Crary, *Suspensions of Perception: Attention, Spectacle and Modern Culture* (Cambridge, MA: MIT Press, 1999), p. 78.

22. Lord Salisbury's speech reported by the *Electrician*, November 8, 1889. Reprinted in Kern, *Culture of Time and Space*, p. 68.

23. Ithiel de Sola Pool, *The Social Impact of the Telephone* (Cambridge, MA: MIT Press, 1977), p. 24.

24. Pierre Teilhard de Chardin, *Toward the Future*, 1st American ed., trans. René Hague (New York: Harcourt Brace Jovanovich, 1975), p. 213. See also Thomas King and James Salmon, ed., *Teilhard and the Unity of Knowledge: The Georgetown University Centennial Symposium* (New York: Paulist Press, 1983). Also James Bix, *Pierre Teilhard de Chardin's Philosophy of Evolution* (Springfield, IL: Thomas, 1972), pp. 5, 11–16, 139.

25. "Measuring the Blogosphere," *New York Times*, August 5, 2005, p. A14.

26. Jon Gertner, "Social Networks," *New York Times Magazine*, December 14, 2003, p. 92.

27. August Fuhrmann, *Das Kaiserpanorama und das Welt-Archiv polychromer Stereo-Urkunden auf Glas* (Berlin, 1905), p. 8. Reprinted in Stephan Oettermann, *The Panorama: History of a Mass Medium*, trans. Deborah Lucas (New York: Zove Books, 1997), p. 230.

28. Benjamin quoted ibid.

29. Angela Miller, "The Panorama, the Cinema and the Emergence of the Spectacle," *Wide Angle* 18, no. 2 (1996): 34–69, here, p. 48.

30. Henry Maudsley, *The Physiology of Mind* (New York: D. Appleton & Co., 1877), p. 310.

31. Michael Hagner, "Toward a History of Attention in Culture and Science," *MLN* 118, no. 3 (2003): 670–87, here, p. 680.

32. Friedrich Nietzsche, "Menschlisches, Allzumenschliches," *Kritische Studienausgabe* (Munich: Deutscher Taschenbuch Verlag, 1980), 2:231. Reprinted in Hagner, "Toward a History of Attention," p. 683.

33. Kern, *Culture of Time and Space*, p. 149.

34. Lorraine Daston, "Attention and the Values of Nature in the Enlightenment," in *The Moral Authority of Nature*, ed. Lorraine Daston and F. Vidal (Chicago: University of Chicago Press, 2004), pp. 100–26.

35. Hagner, "Toward a History of Attention," p. 686.

36. Ibid., p. 679.

37. Bram Stoker, *Dracula: A Norton Critical Edition*, ed. Nina Auerbach and David Skal (New York: W. W. Norton, 1977), p. 252.

38. Crary, *Suspensions of Perception*, p. 69.

39. Joseph Urgo, *In the Age of Distraction* (Jackson: University Press of Mississippi, 2000), p. 169.

40. Sven Birkerts, *The Gutenberg Elegies: The Fate of Reading in an Electronic Age* (Boston: Faber and Faber, 1994), pp. 74–75.

41. Freud quoted in Ernest Jones, *The Life and Work of Sigmund Freud* (New York: Basic Books, 1953–1957), 2:36–37. Reprinted in Crary, *Suspensions of Perception*, p. 363.

42. Jeremy Rifkin, *The European Dream: How Europe's Vision of the Future Is Quietly Eclipsing the American Dream* (New York: Jeremy P. Tarcher/Penguin, 2004), p. 89.

43. Robert Hendrick, "Albert Robida's Imperfect Future," *History Today* 48, no. 7 (July 1998): 27.

44. Philippe Willems, introduction and critical materials for Albert Robida, *The Twentieth Century*, trans. Philippe Willems (Middletown, CT: Wesleyan University Press, 2004), p. xiii.

45. Ibid., p. 391.

CHAPTER TWO

1. Interview with Alan Edelson, May 2006.

2. David Kesmodel, "To Find a Mate, Raid a Dungeon or Speak Like an Elf," *Wall Street Journal*, June 9, 2006, p. A1.

3. Manuel Castells, *The Rise of the Networked Society*, 2nd ed. (Malden, MA: Blackwell Publishers, 2000), 1:403–404.

4. Sherry Turkle, "Computer Games as Evocative Objects: From Projective Screens to Relational Artifacts," in *Handbook of Computer Games Studies*, ed. Joost Raessens and Jeffrey Goldstein (Cambridge, MA: MIT Press, 2005), p. 278.

5. Barry Wellman, "Changing Connectivity: A Future History of Y2.03K," *Sociological Research Online* 4, no. 4 (2000): sect. 1.5, http://www.socresonline.org.uk/4/4/wellman.html.

6. William Gibson, *Neuromancer* (New York: Ace Books, 1984), p. 4.

7. Quoted in Pamela Roberts, "The Living and the Dead: Community in the Virtual Cemetery," *Omega: The Journal of Death & Dying* 49, no. 1 (2004): 57–76, here, pp. 60–61.

8. Interview with Pamela Roberts, May 2006.

9. Warren St. John, "Rituals of Grief Go Online as Web Sites Set Up to Celebrate Life Recall Lives Lost," *New York Times*, April 27, 2006, p. A19.

10. Sandra Gilbert, *Death's Door: Modern Dying and the Ways We Grieve* (New York: W. W. Norton, 2006), p. 247.

11. David Wendell Moller, *Confronting Death* (New York: Oxford University Press, 1996), p. 134.

12. St. John, "Rituals of Grief."

13. Gilbert, *Death's Door*, p. 84.

14. Pamela Roberts and Deborah Schall, "'Hey Dad It's Me Again . . .': Visiting in the Cyberspace Cemetery," paper presented at the Seventh Death, Dying, and Disposal Conference, Bath, UK, September 15–19, 2005.

15. Dagobert Runes, ed., *The Diary and Sundry Observations of Thomas Alva Edison* (New York: Philosophical Library, 1948), p. 233.

16. Michael Benedikt, ed., *Cyberspace: First Steps* (Cambridge, MA: MIT Press, 1991), p. 131.

17. Margaret Wertheim, *The Pearly Gates of Cyberspace: A History of Space from Dante to the Internet* (New York: W. W. Norton, 1999), pp. 21–23.

18. William Shakespeare, *The Tragedy of Hamlet, Prince of Denmark*, act III, scene 1, 79.

19. Yi-Fu Tuan, *Space and Place: The Perspective of Experience* (Minneapolis: University of Minnesota Press, 1977), pp. 6, 52, 54, 140.

20. Randal Walser, "Elements of a Cyberspace Playhouse," in *Virtual Reality: Theory, Practice and Promise*, ed. S. K. Helsel and J. Paris Roth (Westport: Meckler, 1991), p. 53. Reprinted in Sarah Waters, "Ghosting the Interface: Cyberspace and Spiritualism," *Science as Culture* 6, no. 3 (1997): 415.

21. Hans Moravec, *Robot: Mere Machine to Transcendent Mind* (New York: Oxford University Press, 1999), p. 167.

22. Stef Aupers, "The Revenge of the Machines: On Modernity, Digital Technology and Animism," *Asian Journal of Social Science* 30, no. 2 (2002): 199–220, here, p. 216.

23. Interviews with Mae, Beth, and Willie Cohen, May 2006.

24. Barry Wellman, "Changing Connectivity," sects. 7.2–7.3, 8.2.

25. Jeffrey Boase and Barry Wellman, "Personal Relations On and Off the Internet," in *The Cambridge Handbook of Personal Relationships*, ed. Daniel Perlman and Anita L. Vangelisti (Cambridge: Cambridge University Press, 2006), pp. 709–23.

26. danah boyd, Jeff Potter, and Fernanda Viegas, "Fragmentation of Identity through Structural Holes in Email," paper presented at International Sunbelt Social Network Conference XXII, New Orleans, February 13–17, 2002.

27. Interview with Clay Shirky, May 2006.

28. Michael Erard, "Decoding the New Cues in Online Society," *New York Times*, November 27, 2003, p. G1.

29. John Perry Barlow quoted in Paul Tough, "What Are We Doing Online?" *Harper's Magazine*, August 1995, pp. 35–46.

30. E. M. Forster, "The Machine Stops," in *The Machine Stops and Other Stories*, ed. Rod Mengham (London: Andre Deutsch, 1997), p. 87.

31. Robert Kraut et al., "Internet Paradox Revisited," *Journal of Social Issues* 58, no. 1 (2002): 49–74, here, pp. 61, 67–69. Also Irina Shklovski, Robert Kraut, and Lee Rainie, "The Internet and Social Participation: Contrasting Cross-Sectional and Longitudinal Analyses," *Journal of Computer Mediated Communication* 10, no. 1 (2004), http://jcmc.Indiana.edu/vol10/issue1/.

32. Ariana E. Cha, "Home Alone," *Washington Post*, July 13, 2003, p. W8.

33. Edward Castronova, "Virtual Worlds: A First-Hand Account of Market and Society on the Cyberian Frontier," *Gruter Institute Working Papers on Law, Economics and Evolutionary Biology* 2, no. 1 (2001): 1–66, http://www.bepress.com/giwp/default/vol2/iss1/art1.

34. Yi-Fu Tuan, *Escapism* (Baltimore, MD: Johns Hopkins University Press, 1998), p. xvi.

35. Wellman, *Changing Connectivity*, sect. 7.12.

36. Boase and Wellman, "Personal Relations."

37. Jennifer Egan, "Love in the Time of No Time," *New York Times Magazine*, November 23, 2003, p. 66.

38. Interview with Miguel de los Santos, May 2006.

39. Jan-Willem Huisman and Hanne Marckmann, "I Am What I Play: Participation and Reality as Content," in *The Handbook of Computer Game Studies*, ed. Joost Raessens and Jeffrey Goldstein (Cambridge, MA: MIT Press, 2005), p. 397.

40. Barry Wellman and Bernie Hogan, "The Immanent Internet," in *Netting Citizens: Exploring Citizenship in the Internet Age*, ed. Johnston McKay (St. Andrews, Scotland: University of St. Andrews Press, 2004), pp. 54–80.

41. Jeffrey Boase et al., "The Strength of Internet Ties" (Washington, DC: Pew Internet and American Life Project, 2006), pp. vi, 16, http://www.pewinternet.org/pdfs/PIP_Internet_ties.pdf.

42. Miller McPherson, Matthew Brashears, and Lynn Smith-Lovin, "Social Isolation in America: Changes in Core Discussion Networks over Two Decades," *American Sociological Review* 71 (2006): 353–75, here, p. 358.

43. Amanda Lenhart, "Teens and Technology: Youth Are Leading the Transition to a Fully Wired and Mobile Nation" (Washington, DC: Pew Internet & American Life Project, July 27, 2005), pp. iv, 13.

44. Interview with Elinor Ochs, May 2006. See also Joseph Verrengia, "American Families' Plight: Lives Structured to a Fault," *Seattle Times*, March 20, 2005, p. A3.

45. Alessandro Duranti, "Universal and Culture-Specific Properties of Greetings," *Journal of Linguistic Anthropology* 7, no. 1 (1997): 63–97, here, pp. 67–68. Also interview with Duranti, May 2006.

46. Candace West, "Social Accessibility and Involvement: Challenges of the Twenty-first Century," *Contemporary Sociology* 29, no. 4 (2000): 584–90.

47. Ellen Galinsky et al., *Overwork in America: When the Way We Work Becomes Too Much* (New York: Families and Work Institute, 2005), p. 5.

48. Interviews with Tammy, Randy, Jordan, and Lindsay Browning, June 2006.

49. Anthony Giddens, *The Transformation of Intimacy: Sexuality, Love and Eroticism in Modern Societies* (Cambridge, MA: Polity Press, 1992), p. 96.

50. Michael Heim, "The Erotic Ontology of Cyberspace," in *Reading Digital Trend*, ed. David Trend (Oxford: Blackwell Publishers, 2001), p. 81.

51. Paul Virilio, "Speed and Information: Cyberspace Alarm!" ibid., p. 24.

52. Kate Zernike, "First, Your Water Was Filtered. Now It's Your Life," *New York Times*, March 21, 2004, p. WK4.

53. Albert Borgmann, *Holding on to Reality: The Nature of Information at the Turn of the Millennium* (Chicago: University of Chicago Press, 1999), pp. 22, 106, 110.

54. E. B. White, *One Man's Meat* (Gardiner, ME: Tilbury House, 1997, 1938), p. 3.

55. Mark Slouka, *War of the Worlds: Cyberspace and the High-Tech Assault on Reality* (New York: Basic Books, 1995), p. 3.

56. Sherry Turkle, *The Second Self: Computers and the Human Spirit*, 20th anniversary ed. (Cambridge, MA: MIT Press, 2005), p. 15.

57. Borgmann, *Holding on to Reality*, p. 185.

58. Nick Bunkley and Micheline Maynard, "Hoffa Search Finds Town's Sense of Humor," *New York Times*, May 24, 2006, p. A21.

59. Thomas Lynch, *Booking Passage: We Irish and Americans* (New York: W. W. Norton, 2005). Interview with Lynch, May 2006.

60. W. H. Auden, "Musée de Beaux Arts," in *An Introduction to Poetry*, 11th ed., ed. X. J. Kennedy and Dana Gioia (New York: Pearson Longman, 2005), p. 450.

CHAPTER THREE

1. Interview with Daniel Anderson, May 2006.

2. Daniel Anderson and Heather Kirkorian, "Attention and Television," in *The Psychology of Entertainment*, ed. J. Bryant and P. Vorderer (Mahway, NJ: Lawrence Erlbaum, 2006), pp. 35–54.

3. John E. Richards and Daniel Anderson, "Attentional Inertia in Children's Extended Looking at Television," *Advances in Child Development and Behavior*, ed. R. V. Kail (Amsterdam: Academic Press, 2004), 32: 163–212, here, p. 168.

4. Daniel Anderson and Tiffany Pempek, "Television and Very Young Children," *American Behavioral Scientist* 48, no. 5 (January 2005): 505–22, here, p. 508.

5. Marie Schmidt et al., "The Effects of Background Television on the Toy Play of Very Young Children," *Child Development*, in press.

6. Heather Kirkorian et al., "The Impact of Background Television on Parent-Child Interaction," poster presented at the biannual meeting of the Society for Research in Child Development, Atlanta, April 2005.

7. Victoria Rideout and Donald Roberts, *Generation M: Media in the Lives of Eight- to 18-Year-Olds* (Menlo Park, CA: Henry J. Kaiser Family Foundation, March 2005), p. 9.

8. Marshall McLuhan, *Understanding Me: Lectures and Interviews*, ed. Stephanie McLuhan and David Staines (Cambridge, MA: MIT Press, 2003), p. 129.

9. Barbara Schneider and N. Broege, "Why Working Families Avoid Flexibility: The Costs of Over Working," paper presented at the Alfred P. Sloan International Conference "Why Workplace Flexibility Matters," Chicago, May 17, 2006.

10. Eulynn Shiu and Amanda Lenhart, "How Americans Use Instant Messaging" (Washington, DC: Pew Internet & American Life Project, 2004), http://www.pewinternet.org/PPF/r/133/report_display.asp.

11. Bonka Boneva et al., "Teenage Communication in the Instant Messaging Era," in *Computers, Phones and the Internet: Domesticating Information Technology*, ed. Robert Kraut, Malcolm Bryin, and Sara Kiesler (Oxford: Oxford University Press, 2006), pp. 201–18.

12. Lisa Guernsey, "In the Lecture Hall, A Geek Chorus," *New York Times*, July 24, 2003, Circuits sect., p. 1.

13. Ibid.

14. Caryn James, "Splitting. Screens. For Minds. Divided," *New York Times*, January 9, 2004, p. E1.

15. August Fuhrmann, *Das Kaiserpanorama und das Welt-Archiv polychromer Stereo-Urkunden auf Glas* (Berlin, 1905), p. 8. Reprinted in Stephan Oettermann, *The Panorama: History of a Mass Medium*, trans. Deborah Lucas Schneider (New York: Zone Books, 1997), p. 230.

16. Interview with David Meyer, May 2006.

17. Jonathan Crary, *Suspensions of Perception: Attention, Spectacle and Modern Culture* (Cambridge, MA: MIT Press, 1999), p. 29.

18. Ibid., pp. 11–12, 27.

19. Arthur Jersild, "Mental Set and Shift," *Archives of Psychology* 29 (1927).

20. Interviews with Steven Yantis and David Meyer. May, June, and July 2006.

21. David E. Meyer, *Professional Biography Published on the Occasion of His Distinguished Scientific Contribution Award* (Washington, DC: American Psychological Association, 2002), http://www.umich.edu/~bcalab/Meyer_Biography.html.

22. John Serences and Steven Yantis, "Selective Visual Attention and Perceptual Coherence," *Trends in Cognitive Sciences* 10, no. 1 (2006): 38–45. Also Steven Yantis, "How Visual Salience Wins the Battle for Awareness," *Nature Neuroscience* 8, no. 8 (2005): 975–77.

23. Serences and Yantis, "Selective Visual Attention," p. 43.

24. Yantis, "How Visual Salience Wins," p. 975.

25. Susan Landry et al., "Early Maternal and Child Influences on Children's Later Independent Cognitive and Social Functioning," *Child Development* 71, no. 2 (2000): 358–75, here, p. 370.

26. Charles O'Connor, Howard Egeth, and Steven Yantis, "Visual Attention: Bottom-Up versus Top-Down," *Current Biology* 14, no. 19 (2004): R850–52.

27. "Linda Stone's Thoughts on Attention," http://continuouspartial attention.jot.com/WikiHome.

28. Alan Lightman, "The World Is Too Much with Me," in *Living with the Genie: Essays on Technology and the Quest for Human Mastery*, ed. Alan Lightman, Daniel Sarewitz, and Christine Dresser (Washington, DC: Island Press, 2003), pp. 287–303, here, pp. 287, 292.

29. Joshua Rubenstein, David Meyer, and Jeffrey Evans, "Executive Control of Cognitive Processes in Task-Switching," *Journal of Experimental Psychology, Human Perception and Performance* 27, no. 4 (2001): 763–97.

30. Peter F. Drucker, *Age of Discontinuity* (New York: Harper & Row, 1969), p. 271. Also Peter Drucker, "The Coming Rediscovery of Scientific Management," *Conference Board Record* 13 (June 1976): 13–27, here, p. 26.

31. Daniel A. Wren and Ronald Greenwood, *Management Innovators: The People and Ideas That Have Shaped Modern Business* (New York: Oxford University Press, 1998), pp. 134–35, 139. Also Dilys Robinson, "Management Theorist: Thinkers for the 21st Century?" *Training Journal* (January 2005): 30–32.

32. Drucker, "Coming Rediscovery of Scientific Management."

33. Wren and Greenwood, *Management Innovators*, p. 138.

34. Charles Wrege and Ronald Greenwood, *Frederick W. Taylor: The Father of Scientific Management: Myth and Reality* (Homewood, IL: Business One Irwin, 1991), p. 254.

35. Daniel Nelson, "Frederick W. Taylor," American National Biography Online, http://www.anb.org/articles/20/20-01725.html.

36. Wren and Greenwood, *Management Innovators*, p. 136.

37. Walter Benjamin, "The Age of Mechanical Reproduction," in *Illuminations*, ed. Hannah Arendt (New York: Harcourt, Brace & World, 1968), p. 237.

38. Charles Leland, "Quickness of Perception," *Memory and Thought*, vol. 2 (Harrisburg, PA: J. P. Downs, 1891). Reprinted in Stephen Arata, "On Not Paying Attention," *Victorian Studies* 46, no. 2 (Winter 2004): 193–205, here, p. 199.

39. Ibid.

40. Nelson, "Frederick W. Taylor."

41. Merle Thomas, "The Gold Standard," *Industrial Engineer* 38, no. 4 (2006): 35.

42. Peter Drucker, *Management Challenges for the 21st Century* (New York: Harper Business, 1999), p. 138.

43. "Report of a Lecture by and Questions Put to Mr. F. W. Taylor," *Journal of Management History* 1, no. 1 (1995): 8–32, here, p. 10. Also Robinson, "Management Theorist."

44. Wren and Greenwood, *Management Innovators*, p. 143.

45. Clive Thompson, "Meet the Life Hackers," *New York Times Magazine*, October 16, 2005, pp. 40–45.

46. Gloria Mark, Victor Gonzalez, and Justin Harris, "No Task Left Behind? Examining the Nature of Fragmented Work," proceedings of the Conference on Human Factors in Computer Systems (Portland, OR, 2005): 321–30. Also interview with Gloria Mark, July 2006.

47. Ibid.

48. Thompson, "Meet the Life Hackers," p. 42.

49. Tony Gillie and Donald Broadbent, "What Makes Interruptions Disruptive? A Study of Length, Similarity and Complexity," *Psychological Research* 50 (1989): 243–50.

50. Jonathan Spira and Joshua Feintuch, *The Cost of Not Paying Attention: How Interruptions Impact Knowledge Worker Productivity* (New York: Basex, 2005), pp. 2, 10.

51. Mark, Gonzalez, and Harris, "No Task Left Behind," p. 326.

52. Thompson, "Meet the Life Hackers."

53. Mark, "No Task Left Behind," p. 324. Also ibid.

54. Henry James, "A Novelist's View of the Morrises," in *William*

Morris: The Critical Heritage, ed. Peter Faulkner (London: Routledge Kegan Paul, 1973), pp. 77–78.

55. Peter Ackroyd, "Blooming Genius," *New Yorker*, September 23, 1996, pp. 90–94.

56. Edward Carpenter, "Morris' 'Great Inspiring Hatred,'" in Faulkner, *William Morris*, pp. 401–403.

57. Ackroyd, "Blooming Genius."

58. Ibid.

59. Peter Stansky, "Morris," in *Victorian Thinkers: Carlyle, Ruskin, Arnold, Morris* (Oxford: Oxford University Press, 1993), pp. 327–408, here, p. 345.

60. William Morris, "The Revival of Handicraft," *Fortnightly Review* 1 (November 1888): 603–11, here, p. 606.

61. William Morris, *News from Nowhere: Or an Epoch of Rest, Being Some Chapters from a Utopian Romance*, ed. Krishan Kumar (Cambridge: Cambridge University Press, 1995), p. 213.

62. Arata, "On Not Paying Attention," p. 199.

63. William Butler Yeats, "Review," in *Bookman* (November 1896): x, 37–38. Reprinted in Faulkner, *William Morris*, pp. 415–17. Also George Bernard Shaw, *Morris as I Knew Him* (New York: Dodd, 1936), p. 41.

64. Yeats, "Review."

65. Suzanne Ross, "Two Screens Are Better Than One," *Microsoft Research News and Highlights*, http://research.microsoft.com/displayArticle.aspx?id=433&0sr=a. Also Tara Matthews et al., "Clipping Lists and Change Borders: Improving Multitasking Efficiency with Peripheral Information Design," *Proceedings of the Conference on Human Factors in Computer Systems* (April 2006): 989–98.

66. Scott Brown and Fergus I. M. Craik, "Encoding and Retrieval of Information," *Oxford Handbook of Memory*, ed. Endel Tulving and Fergus I. M. Craik (New York: Oxford University Press, 2000), pp. 93–107, here, p. 79.

67. Ibid. Also Sadie Dingfelder, "A Workout for Working Memory," *Monitor on Psychology* 36, no. 8 (2005), http://www.apa.org/monitor/sep05/workout.html. Jan de Fockert et al., "The Role of Working Memory in Visual Selective Attention," *Science* 291, no. 5509 (2001): 1803–1804.

68. Lori Bergen, Tom Grimes, and Deborah Potter, "How Attention Partitions Itself during Simultaneous Message Presentations," *Human Communication Research* 31, no. 3 (2005): 311–36.

69. Interview with Mary Czerwinski, July 2006.

70. W. Wayt Gibbs, "Considerate Computing," *Scientific American*, January 2005, pp. 55–61. Also Peter Weiss, "Minding Your Business," *Science News* 163, no. 18 (2006): 279.

71. Horwitz quoted in Gibbs, "Considerate Computing."

72. Searle quoted in Weiss, "Minding Your Business."

73. Paul Virilio, *The Vision Machine* (Bloomington: Indiana University Press, 1994), p. 59.

74. Jane Healy, *Endangered Minds: Why Our Children Don't Think* (New York: Simon & Schuster, 1990), p. 153.

75. Arthur T. Jersild, "Reminiscences of Arthur Thomas Jersild: Oral History 1967," interviewer T. Hogan (New York: Columbia University, 1972), pp. 2, 20, 40–41, 79, 246.

76. Brown and Craik, *Oxford Handbook of Memory*, pp. 93–97. Also John T. Wixted, "A Theory about Why We Forget What We Once Knew," *Current Directions in Psychological Science* 14, no. 1 (2005): 6–9.

77. Alan Lightman, *The Diagnosis* (New York: Pantheon Books, 2000), pp. 3–20.

CHAPTER FOUR

1. Interview with Regina Lewis, September 2006.

2. Kate Murphy, "Look! We Can Drive and Snack at the Same Time," *New York Times*, November 2, 2003, p. 3.

3. Biing-Hwan Lin and Elizabeth Frazao, "Away-from-Home Foods Increasingly Important to Quality of American Diet," *US Department of Agriculture Information Bulletin* 749 (January 1999).

4. Interview with Harry Balzer, vice president of NPD Group Market Research, September 2006.

5. Ibid.

6. Paul Rozin, "The Meaning of Food in Our Lives: A Cross-Cultural Perspective on Eating and Well-Being," *Journal of Nutrition, Education and Behavior* 37, supplement 2 (2005): 19–24.

7. Zygmunt Bauman, *Globalization: The Human Consequences* (New York: Columbia University Press, 1998), p. 2.

8. George Pierson, "A Restless Temper," *American Historical Review* 69, no. 4 (July 1964): 969–89, here, p. 980.

9. Sylvia Hilton and Cornelis van Minnen, *Nation on the Move: Mobility in U.S. History* (Amsterdam: VU University Press, 2002), p. 4.

10. Ibid., p. 5.

11. Alexis de Tocqueville, *Journey to America*, ed. J. P. Mayer (New Haven, CT: Yale University Press, 1960), pp. 182–83. Also Alexis de Tocqueville, *Democracy in America*, ed. Phillips Bradley (New York: Knopf, 1945), 2:136–7. Reprinted in Pierson, "Restless Temper," pp. 987–88.

12. Nigel Thrift, *Spatial Formations* (London: Sage, 1996), p. 266.

13. Jason Schachter, *Geographic Mobility: 2002 to 2003* (Washington, DC: US Census Bureau Current Population Reports, March 2003). Also Douglas Wolf and Charles Longino Jr., "Our 'Increasingly Mobile Society'? The Curious Persistence of a False Belief," *Gerontologist* 45, no. 1 (2005): 5. June Kronholz, "The Coming Crunch," *Wall Street Journal*, October 13, 2006, p. B1.

14. Kenneth Gergen, *The Saturated Self: Dilemmas of Identity in Contemporary Life* (New York: Basic Books, 1991), p. 61.

15. John Urry, "Mobile Cultures," Lancaster (UK) University Department of Sociology, 1999, http://www.lancs.ac.uk/fass/sociology/research/restopic.htm.

16. Cullen Murphy, "The Oasis of Memory," *Atlantic Monthly*, May 1998, p. 24.

17. Bauman, *Globalization*, pp. 121–22.

18. Interview with Jaime Eshak, October 2006.

19. Data provided to author by Mintel Global New Products Database, Mintel International, October 2007.

20. Interview with Mekonnen Kebede, October 2006.

21. "Food Bars: A Small Package That's Leading to Large Profits," *Packaged Facts* (Rockville, MD: Market Research Group, September 2003). Also *Times and Trends* (Chicago: Information Resources, May 2003), p. 17, November 2003, p. 3, and April 2004, p. 6.

22. Interview with Balzer.

23. Interview with Kevin Elliott, October 2006.

24. Clotaire Rapaille, *The Culture Code* (New York: Broadway Books, 2006), pp. 143, 146.

25. Suzanne Bianchi, John P. Robinson, and Melissa Milkie, *Changing Rhythms of American Family Life* (New York: Russell Sage Foundation, 2006), p. 95.

26. Paul Rozin et al., "The Ecology of Eating: Smaller Portion Sizes in

France Than in the United States Help Explain the French Paradox," *Psychological Science* 14, no. 5 (September 2003): 450–54, here, p. 453.

27. Eric Schlosser, *Fast Food Nation* (Boston: Houghton Mifflin, 2001), p. 3.

28. Rapaille, *Culture Code*, pp. 109, 146.

29. Carole Sugarman, "Grab It and Go!: Convenience Foods Take on a Whole New Life," *Washington Post*, January 6, 1999, p. E01.

30. Barbara Bloemink, introduction to Sarah D. Coffin et al., *Feeding Desire: Design and the Tools of the Table* (New York: Assouline Publishing and the Smithsonian Institution, 2006), p. 7.

31. Darra Goldstein, "Implements of Eating," ibid., p. 116.

32. Yi-Fu Tuan, *Segmented Worlds and Self: Group Life and Individual Consciousness* (Minneapolis: University of Minnesota Press, 1982), pp. 40–43.

33. Goldstein, "Implements," pp. 118–19.

34. Tuan, *Segmented Worlds*, pp. 45, 50.

35. "Room with a View: Needing Some Space," *Wilson Quarterly* 30, no. 2 (2006): 12. Also Jamie Horwitz, "Meals in Transit," paper presented to the Alternative Mobility Futures Conference, Lancaster (UK) University, January 9–11, 2004.

36. T. E. Lawrence, *Seven Pillars of Wisdom* (Garden City, NY: Doubleday, 1938). Reprinted in John Ure, *In Search of Nomads: An Anglo-American Obsession from Hester Stanhope to Bruce Chatwin* (London: Constable, 2003), p. 105.

37. Morris Berman, *Wandering God: A Study in Nomadic Spirituality* (Albany: State University of New York Press, 2000), p. 165.

38. Ssu-Ma-Ch'ien, *Records of the Grand Historian of China*, chap. 108, trans. Burton Watson (New York: Columbia University Press, 1961). Reprinted in Bruce Chatwin, *Anatomy of Restlessness: Selected Writings*, ed. Jan Borm and Matthew Graves (London: Cape, 1996), p. 94.

39. Daniel Cohen, *Conquerors on Horseback* (Garden City, NY: Doubleday, 1970), p. 151.

40. Ure, *In Search of Nomads*, p. 222.

41. Chatwin, *Anatomy*, p. 11.

42. Isabelle Eberhardt, *The Nomad: The Diaries of Isabelle Eberhardt*, ed. Elizabeth Kershow (New York: Interlink Books, 2003), pp. 25, 180.

43. David Manners and Tsugio Makimoto, *Digital Nomad* (Chichester, England: Wiley, 1997), p. 21.

44. Fernand Braudel, *The Mediterranean and the Mediterranean World in the Age of Phillip II*, trans. Sian Reynolds (New York: Harper & Row, 1966), 1:100.

45. Owen Lattimore, *Nomads and Commissars: Mongolia Revisited* (New York: Oxford University Press, 1962).

46. Ure, *In Search of Nomads*, pp. 148, 154.

47. Alan Pisarski, *Commuting in America III: The Third National Report on Commuting Patterns and Trends* (Washington, DC: Transportation Research Board, 2006), http://onlinepubs.trb.org/onlinepubs/nchrp/CIAIII.pdf.

48. Peter Wilson, *The Domestication of the Human Species* (New Haven, CT: Yale University Press, 1988), pp. 50, 57.

49. Claire Parnet, "A Conversation: What Is It? What Is It For?" in *Dialogues II*, by Gilles Deleuze and Claire Parnet, trans. Barbara Habberjam and Hugh Tomlinson (New York: Columbia University Press, 2007, rev. ed.; originally published, 1977), p. 31.

50. Pico Iyer, *The Global Soul: Jet Lag, Shopping Malls and the Search for Home* (New York: Knopf, 2000), pp. 18, 19, 24.

51. Bauman, *Globalization*, p. 121.

52. Fleura Bardhi and Eric J. Arnould, "Making a Home on the Road: A Mobile Concept of Home," in *Advances in Consumer Research* 33 (2006): 651–54. Also interview with Fleura Bardhi, September 2006.

53. Chatwin, *Anatomy*, pp. 182–83.

54. Yi-Fu Tuan, *Space and Place: The Perspective of Experience* (Minneapolis: University of Minnesota Press, 1977), pp. 6, 54, 179.

55. Niels Bohrs quoted in Werner Heisenberg, *Physics and Beyond: Encounters and Conversations* (New York: Harper Torchbook, 1972), p. 51. Reprinted ibid., p. 4.

56. Yi-Fu Tuan, "Space and Place: A Humanist Perspective," in *Progress in Geography* (London: Edward Arnold, 1974): 6:211–52, here, p. 241.

57. Yi-Fu Tuan, *Who Am I? An Autobiography of Emotion, Mind and Spirit* (Madison: University of Wisconsin Press, 1999), pp. 4, 94, 130.

58. Denis Wood quoted in "Mapping," *This American Life Episode 110*, September 4, 1998, http://www.thisamericanlife.org/Radio_Episode.aspx?episode=110.

59. *Encyclopedia Britannica* 1910 quoted in Stephen Hall, *Mapping the Next Millennium: The Discovery of New Geographies* (New York: Random House, 1992), p. 4.

60. Interview with Margaret Pearce, September 2006.

61. Cohen, *Conquerors*, pp. 66, 161.

62. Wilson, *The Domestication of the Human Species*, p. 50.

63. Interview with Mushon Zer-Aviv, September 2006.

64. Interview with Denis Wood, September 2006.

65. Rosalind Williams, *Noted on the Underground: An Essay on Technology, Society and the Imagination* (Cambridge, MA: MIT Press, 1990), p. 2.

66. John Urry, *Sociology beyond Societies: Mobilities for the Twenty-first Century* (London: Routledge, 2000), p. 63.

67. Michael Bull, *Sounding Out the City: Personal Stereos and the Management of Everyday Life* (Oxford: Berg, 2000), p. 41.

68. Bill McKibben, *The Age of Missing Information* (New York: Random House, 1992), p. 9.

69. Noelle Oxenhandler, "The Lost While," in *How We Want to Live: Narratives of Progress*, ed. Susan Richards Shreve and Porter Shreve (London: Routledge, 2000), p. 92.

70. Tuan, *Segmented Worlds*, p. 115.

71. Karen Hamrick and Kristina J. Shelley, "How Much Time Do Americans Spend Preparing and Eating Food?" *Amber Waves* (November 2005): 1–4, http://www.ers.usda.gov//AmberWaves/November05/DataFeature. Also Rozin, "The Meaning of Food in Our Lives," pp. S21–S22.

72. Interview with Carol Devine, September 2006.

73. Michael Pollan, *The Omnivore's Dilemma: A Natural History of Four Meals* (New York: Penguin Press, 2006), pp. 111–12.

74. Interview with Brian Wansink, September 2006.

75. Marina Warner, "The Word Unfleshed: Memory in Cyberspace," *Raritan* 26, no. 1 (2006): 1.

CHAPTER FIVE

1. C. F. A. Marmoy, "The 'Auto-Icon' of Jeremy Bentham at University College, London," *Medical History* 2 (1958): 77–86. Also "Jeremy Bentham," University College London Bentham Project, http://www.ucl.ac.uk/Bentham-Project.

2. Kelly McCollum, "Founder of Utilitarianism Is Present in Spirit at 250th Birthday Teleconference," *Chronicle of Higher Education* 44, no. 25 (1998): A28.

3. Catherine Pease-Watkin, "Bentham's Panopticon and Dumont's *Panoptique*," University College London Bentham Project (2003), http://www.ucl.ac.uk/Bentham-Project/journal/cpwpan.htm.

4. David Lyon, *Surveillance Society: Monitoring Everyday Life* (Buckingham: Open University Press, 2001), p. 15.

5. Jeremy Bentham, *The Correspondence of Jeremy Bentham*, vol. 4, ed. Alexander Milne (1788–1793), p. 219.

6. Jeremy Bentham, *The Works of Jeremy Bentham*, vol. 4 (New York: Russell & Russell, 1962), pp. 81–82.

7. Interview with "Jim," November 2006.

8. Anthony Giddens, *Conversations with Anthony Giddens: Making Sense of Modernity* (Cambridge: Polity Press, 1998), pp. 102–104.

9. Interview with Steven Mintz, December 2006.

10. Interview with Dan Pope, December 2006.

11. Steven Flusty, "Building Paranoia," in *The Architecture of Fear*, ed. Nan Ellin (New York: Princeton Architectural Press, 1997), pp. 46–48.

12. Michel Foucault, *Psychiatric Power: Lectures at the College de France, 1973–74*, ed. Jacques Lagrange, trans. Graham Burchell (Basingstoke, England: Palgrave Macmillan, 2006), pp. 41–47.

13. Michel Foucault, *Discipline and Punish: The Birth of the Prison*, trans. Alan Sheridan (New York: Random House, 1995), p. 201.

14. Interview with Peter Kleiner, December 2006.

15. Larry Selditz, Road Safety International Web site, http://www.roadsafety.com/pressroom.php.

16. Lyon, *Surveillance Society*, p. 60.

17. Diane Ackerman, *A Natural History of the Senses* (New York: Random House, 1990), pp. 229–30.

18. Yi-Fu Tuan, *Segmented Worlds and Self: Group Life and Individual Consciousness* (Minneapolis: University of Minnesota Press, 1982), pp. 118–34.

19. John Urry, *Sociology beyond Societies: Mobilities for the Twenty-first Century* (London: Routledge, 2000), p. 25.

20. Jason McCarley, "Elements of Human Performance in Baggage X-Ray Screening," paper presented at the Fourth International Aviation Security Technology Symposium, Washington, DC, November 29, 2006.

21. Jeremy Wolfe, Todd Horowitz, and M. J. Van Wert, "The Prevalence Problem in Visual Search," paper presented at the Fourth International Aviation Security Technology Symposium, Washington, DC, November 29, 2006.

22. Daniel J. Simons and Christopher F. Chabris, "Gorillas in Our Midst: Sustained Inattentional Blindness for Dynamic Events," *Perception* 28 (1999): 1059–74.

23. James Gorman, "Come Here Often? And by the Way, Did You Happen to Notice That Gorilla?" *New York Times*, July 11, 2006, p. F3. Seema Clifasefi, Melanie Takarangi, and Jonah Bergman, "Blind Drunk: The Effects of Alcohol on Inattentional Blindness," *Applied Cognitive Psychology* 20 (2006): 697–704.

24. Interview with Daniel Simons, January 2007.

25. Jason McCarley et al., "Conversation Disrupts Change Detection in Complex Traffic Scenes," *Human Factors* 46, no. 3 (2004): 424–36.

26. Walter R. Boot et al., "Detecting Transient Changes in Dynamic Displays: The More You Look, the Less You See," *Human Factors* 48 (2006): 759–73.

27. Michael Shermer, "None So Blind," *Scientific American*, March 2004, p. 42.

28. Ackerman, *A Natural History of the Senses*, p. 304.

29. Georgia O'Keeffe quoted ibid., p. 267.

30. Madge quoted in Caleb Crain, "Surveillance Society: A Critic at Large," *New Yorker*, September 11, 2006, p. 76.

31. Madge and Jennings quoted ibid.

32. Quoted ibid.

33. Angus Calder and Dorothy Sheridan, *Speak for Yourself: A Mass-Observation Anthology, 1937–49* (London: Jonathan Cape, 1984), pp. 3–17, 153.

34. Marion Dewhirst, "They May Be Watching You," *Everybody*, June 18, 1938. Reprinted ibid., p. 17.

35. Ackerman, *Natural History*, p. 231.

36. Susan Sontag, *On Photography* (New York: Anchor Books, 1990; originally published New York: Farrar, Straus and Giroux, 1977), pp. 3, 5–6.

37. Ibid., pp. 23, 82.

38. Quoted in Crain, "Surveillance Society."

39. Interview with Alice Byrne, December 2006.

40. Robert O'Harrow Jr., *No Place to Hide* (New York: Free Press, 2005), pp. 162–66.

41. Steve Secklow, "Watch on the Thames," *Wall Street Journal*, July 8, 2005, p. B1. Also Libby Sander, "A Tempest When Art Becomes Surveillance," *New York Times*, December 28, 2006, p. A22.

42. O'Harrow, *No Place to Hide*, pp. 230–31.

43. Lyon, *Surveillance Society*, p. 86.

44. Ibid., p. 113.

45. Crain, "Surveillance Society."

46. Sophie Calle, "The Address Book," in *Sophie Calle, M'as Tue Vue*, ed. Christine Macel, trans. Simon Pleasance and Charles Penwarden (Munich: Prestel, 2003), pp. 97–100.

47. Ibid., p. 41.

48. Alan Riding, "Keeping It Together by Living in Public," *New York Times*, December 7, 2003, p. 2. Also Alan Riding, "Intimacy and Strangers Structure Her Life," *New York Times*, April 28, 1999, p. E1.

49. Baudrillard, *Please Follow Me* (Seattle, WA: Bay Press, 1988). Reprinted in Janet Hand, "Sophie Calle's Art of Following and Seduction," *Cultural Geographies* 12 (2005): 463–84, here, p. 479.

50. Russell Hardin, *Trust & Trustworthiness* (New York: Russell Sage Foundation, 2002), p. 12.

51. Trudy Govier, *Social Trust and Human Communities* (London: McGill-Queen's University Press, 1997), p. 4.

52. Hardin, *Trust & Trustworthiness*, pp. 39, 91, 199. Russell Hardin, *Trust* (Cambridge: Polity Press, 2006), p. 18.

53. Interview with Russell Hardin, November 2006.

54. Hardin, *Trust & Trustworthiness*, p. 92.

55. Govier, *Social Trust*, p. 138. See Edward Banfield, *The Moral Basis of a Backward Society* (Glencoe, IL: Free Press, 1958).

56. Interview with Wendy Mogel, December 2006.

57. Hardin, *Trust & Trustworthiness*, pp. 102, 113.

58. Erving Goffman, *Behavior in Public Places: Notes on the Social Organization of Gatherings* (New York: Free Press of Glencoe, 1963), pp. 16, 83.

59. Philip Manning, *Erving Goffman and Modern Sociology* (Stanford, CA: Stanford University Press, 1992), p. 5.

60. Goffman, *Behavior*, pp. 43–44, 246.

61. Giddens, *Conversations with Anthony Giddens*, pp. 101, 125.

62. Riding, "Keeping It Together."

63. William Hazlitt, *The Spirit of the Age* (Oxford: Woodstock Books, 1989), pp. 5, 25.

64. Charles Bahmueller quoted in *The National Charity Company:*

Jeremy Bentham's Silent Revolution (Berkeley: University of California Press, 1981), p. 213.

65. Phillip Lucas and Anne Sheeran, "Asperger's Syndrome and the Eccentricity and Genius of Jeremy Bentham," *Journal of Bentham Studies* 8 (2006): 1–20, here, p. 19.

CHAPTER SIX

1. Adrian Johns, *The Nature of the Book: Print and Knowledge in the Making* (Chicago: University of Chicago Press, 1998), p. 87.

2. Ibid., p. 66.

3. Interview with John Bidwell, February 2007.

4. Andrew Grabois, "Book Title Output and Average Prices: 2005 Final and 2006 Preliminary Figures," in *The Bowker Annual: Library and Book Trade Almanac 2007* (Medford, NJ: Information Today, 2007), p. 487.

5. Mark Kutner et al., *Literacy in Everyday Life: Results from the 2003 National Assessment of Adult Literacy* (Washington, DC: US Department of Education National Center for Education Statistics, 2007). Also *Reading at Risk: A Survey of Literary Reading in America* (Washington, DC: National Endowment for the Arts, 2004), p. ix, http://www.nea.gov/pub/ReadingAtRisk.pdf.

6. William Mitchell, *City of Bits: Space, Place and the Infobahn* (Cambridge, MA: MIT Press, 1995), p. 56.

7. Richard Lanham, *The Electronic Word: Democracy, Technology and the Arts* (Chicago: University of Chicago Press, 1993), pp. 73–74, 243, 253–54.

8. James J. O'Donnell, "The Pragmatics of the New: Trithemius, McLuhan, Cassiodorus," in Geoffrey Nunberg, *Future of the Book* (Berkeley: University of California Press, 1996), p. 54.

9. Marinetti quoted in Luciano de Maria, ed., *Marinetti e il Futurismo* (Milan: Mondadori, 1973), pp. 189–90. Reprinted and translated in Lanham, *Electronic Word*, p. 33.

10. Luce Marinetti, "Notes on Free-Word-Composition," in *Words-in-Freedom: Drawings by the Italian Futurists*, ed. Anne Coffin Hanson (New York: Ex Libris, Prakapas Gallery, 1985).

11. Richard Lanham, *The Economics of Attention: Style and Substance in the Age of Information* (Chicago: University of Chicago Press, 2006), p. 45.

12. Ibid.

13. Plato, *Phaedrus*, trans. C. J. Rowe (Warminster: Aris and Phillips, 1986), p. 125.

14. Raffaele Simone, "The Body of the Text," in Nunberg, *Future of the Book*, p. 245.

15. Johns, *Nature of the Book*, p. 31.

16. Alberto Manguel, *Into the Looking-Glass: Essays on Books, Reading and the World* (San Diego: Harcourt, 2000), p. 268.

17. Robert Darnton, "History of Reading," in *New Perspectives on Historical Writing*, ed. Peter Burke (Cambridge: Polity Press, 1991), pp. 141–67, here, p. 150.

18. Darnton, "History of Reading," p. 150.

19. David Thornburn and Henry Jenkins, introduction to *Rethinking Media Change: The Aesthetics of Transition*, ed. David Thornburn and Henry Jenkins (Cambridge, MA: MIT Press, 2003), pp. 1–16.

20. Harold Bloom, *How to Read and Why* (New York: Scriber, 2000), p. 29.

21. Manguel, *Into the Looking-Glass*, p. 270.

22. Rosalind Thomas, *Literacy and Orality in Ancient Greece* (Cambridge: Cambridge University Press, 1992), p. 4.

23. David Levy, *Scrolling Forward: Making Sense of Documents in the Digital Age* (New York: Arcade Publishing, 2001), pp. 108–109. An influential early work on the fate of print in the digital era.

24. Darnton, "History of Reading," p. 149. Also Roger Chartier, *Forms and Meanings: Texts, Performances, and Audiences from Codex to Computer* (Philadelphia: University of Pennsylvania Press, 1995), pp. 16–17.

25. Christine Pawley, *Reading on the Middle Border: The Culture of Print in Late Nineteenth-Century Osage, Iowa* (Amherst: University of Massachusetts Press, 2001), p. 61.

26. First use of word *literacy* was in 1883. *Oxford English Dictionary* (Oxford: Clarendon Press, 1989), 8:1026.

27. Geoffrey Nunberg, "Farewell to the Information Age," in *Future of the Book*, ed. Nunberg, pp. 116–17.

28. Daniel Boorstin, *Cleopatra's Nose: Essays on the Unexpected*, ed. Ruth Boorstin (New York: Vintage Books Random House, 1995, first published, 1994), pp. 8–10.

29. Walter Ong, *An Ong Reader: Challenges for Further Inquiry*, ed. Thomas Farrell and Paul Soukup (Cresskill, NJ: Hampton Press, 2002), p. 523.

30. Interviews with Norbert Elliot and his students, February 2007.

31. Shawn Lombardo and Cynthia Miree, "Caught in the Web: The Impact of Library Instruction on Business Students' Perceptions and Use of Print and Online Resources," *College and Research Libraries* 64, no. 1 (2003): 6–22, here, p. 30.

32. Deborah Fallows, *Search Engine Users* (Washington, DC: Pew Internet and American Life Project, 2005), pp. iii–iv, 24–27, http://www .pewinternet.org/pdfs/PIP_searchengine_users.pdf.

33. Geoffrey Nunberg, "Teaching Students to Swim in the Online Sea," *New York Times*, February 13, 2005, p. WK4.

34. Patricia Senn Breivik and E. Gordon Gee, *Higher Education in the Internet Age: Libraries Creating a Strategic Edge* (Westport, CT: American Council on Education, Praeger Series on Higher Education, 2006), pp. 10–11.

35. Christen Thompson, "Information Literate or Lazy: How College Students Use the Web for Research," *Libraries and the Academy* 3, no. 2 (2003): 259. Also Nancy Young and Marilyn Von Seggern, "General Information Seeking in Changing Times: A Focus Group Study," *Reference and User Services Quarterly* 41, no. 2 (2001): 159. Also Patti Caravello et al., *Information Competence at UCLA: Report of a Survey Project* (Los Angeles: UCLA Library Instructional Services Advisory Committee, 2001), pp. 1–3, http://www.library.ucla.edu/infocompetence. Hoa Loranger and Jakob Nielsen, *Teenagers on the Web: Usability Guidelines for Creating Compelling Websites for Teens* (Fremont, CA: Nielsen Norman Group, 2005), pp. 5–6, 12–13, 17–18.

36. "2006 Information Computer Training Literacy Assessment Preliminary Findings" (Princeton, NJ: Educational Testing Service, 2006), pp. 1–17.

37. David Scharf, "Direct Assessment of Information Literacy Using Writing Portfolios," *Journal of Academic Librarianship* 33, no. 4 (2007): 462–78.

38. Thomas, *Literacy and Orality*, pp. 10–11.

39. Patricia Davitt Maughan, "Assessing Information Literacy among Undergraduates: A Discussion of the Literature and the University of California/Berkeley Assessment Experience," *College and Research Libraries* 62, no. 1 (2001): 71–85, here, p. 77. Also Thompson, "Information Literate or Lazy."

40. Fallows, *Search Engine Users.*

41. Boorstin, *Cleopatra's Nose*, p. 7.

42. David Crystal, *How Language Works: How Babies Babble, Words Change Meaning, and Languages Live or Die* (Woodstock, NY: Overlook Press, 2006), pp. 121, 133.

43. Interview with Bruce McCandliss, January 2007.

44. Kimberly Noble and Bruce McCandliss, "Reading Development and Impairment: Behavioral, Social and Neurobiological Factors," *Development and Behavioral Pediatrics* 26, no. 5 (2005): 370–78.

45. Ibid., p. 372.

46. Bradley Schlagger and Bruce McCandliss, "Development of Neural Systems for Reading," *Annual Review of Neuroscience* 30, no. 1 (2007): 475–503.

47. Michael Posner and Mary K. Rothbart, *Educating the Human Brain* (Washington, DC: American Psychological Association, 2007), p. 151.

48. Bruce McCandliss, "Cognitive Neuroscience of Reading Development," presentation to Weill Cornell Medical College seminar series for training clinicians in functional neuroimaging, New York City, January 31, 2007.

49. Darnton, "History of Reading," p. 152.

50. Johns, *Nature of the Book*, pp. 392–97, 405–406.

51. Quoted in Darnton, "History of Reading," p. 152.

52. Bloom, *How to Read and Why*, p. 29.

53. Donna Shannon, "Kuhlthau's Information Search Process," *School Library Media Activities Monthly* 19, no. 2 (2002): 19.

54. John Dewey, *Art as Experience* (New York: Minton, Bach and Company, 1934), p. 35.

55. Carol Collier Kuhlthau, *Seeking Meaning: A Process Approach to Library and Information Services* (Norwood, NJ: Ablex Publishing, 1993), p. 3.

56. John Dewey, *How We Think* (Boston: Heath & Co., 1933), p. 108.

57. Crystal, *How Language Works*, p. 128. Bloom, *How to Read and Why*, p. 25.

58. Kuhlthau, *Seeking Meaning*, p. 68.

59. Dewey, *Art as Experience*, p. 35

60. Manguel, *Into the Looking-Glass*, p. 261.

61. George Kennedy, *A New History of Classical Rhetoric* (Princeton, NJ: Princeton University Press, 1994), pp. 3–4.

62. Walter Ong, *Orality and Literacy: The Technologizing of the Word* (London: Methuen, 1982), pp. 34–35.

63. Ibid., p. 58.

64. Walter Benjamin, *Illuminations*, ed. Hannah Arendt (New York: Harcourt, Brace & World, 1968), pp. 83, 88.

65. Johns, *Nature of the Book*, pp. 1–2, 183, 541, 622.

66. Alberto Manguel, *The Library at Night* (Toronto: A. A. Knopf Canada, 2006), p. 85.

67. Stacy Schiff, "Know It All: Annals of Information," *New Yorker*, July 31, 2006, p. 36.

68. Manguel, *Library at Night*, p. 84.

69. Schiff, "Know It All."

70. Kevin Kelly, "Scan this Book!" *New York Times Magazine*, May 14, 2006, p. 42.

71. Nunberg, "Farewell to the Information Age," in *Future of the Book*, p. 117.

72. John Updike, "The End of Authorship," *New York Times Book Review*, June 25, 2006, p. 27.

73. Martin Graff, "Differences in Concept Mapping, Hypertext Architecture and the Analyst-Intuitive Dimension of Cognitive Style," *Educational Psychology* 25, no. 4 (2005): 409–22. Also Sherry Chen, "A Cognitive Model for Non-linear Learning in Hypermedia Programmes," *British Journal of Educational Technology* 33, no. 4 (2002): 449–60. Also T. de Jong and A. van der Hurst, "The Effects of Graphical Overviews on Knowledge Acquisition in Hypertext," *Journal of Computer-Assisted Learning* 18 (2002): 219–31. Also Richard Overbaugh and Shin Yi Lin, "Student Characteristics, Sense of Community and Cognitive Achievement in Web-Based and Lab-Based Learning Environments," *Journal on Research on Technology in Education* 39, no. 2 (2006): 205–23.

74. Manguel, *Library at Night*, pp. 88–89. Jorge Luis Borge, "The Congress," in *The Book of Sand*, trans. Norman Thomas di Giovanni (New York: E. P. Dutton, 1977), pp. 27–49.

75. Breivik and Gee, *Higher Education*, p. 41.

76. James Marcum, "Rethinking Information Literacy," *Library Quarterly* 72, no. 1 (2002): 1.

77. Updike, "End of Authorship."

78. Nunberg "Farewell," in *Future of the Book*, pp. 126–27.

79. Paula Bernstein, "The Book as Place: The 'Networked Book' Becomes New 'In' Destination," *Searcher* 14, no. 10 (November/December 2006), http://

www.infotoday.com/searcher/nov06/Berinstein.shtml. See also McKenzie Wark, *Gamer Theory* (Cambridge, MA: Harvard University Press, 2007).

80. Ong, *Ong Reader*, pp. 546–47.

CHAPTER SEVEN

1. Interview with Aaron Edsinger, May 2007.

2. Edsinger quoted in interview with Bruce Gellerman, "The Future of Robotics," from *Living on Earth* radio show on WBUR Boston, January 12, 2007, http://www.loe.org/shows/segments.htm?programID=07-P13-00002 &segmentID=8.

3. Gaby Wood, *Edison's Eve: A Magical History of the Quest for Mechanical Life* (New York: Knopf, 2002), pp. 21, 60.

4. Nigel Thrift, *Spatial Formations* (London: Sage, 1996), p. 279.

5. Sherry Turkle, "Diary," *London Review of Books*, April 20, 2006, pp. 36–37.

6. Walter Dan Stiehl et al., "The Huggable: A Therapeutic Robotic Companion for Relational, Affective Touch," demonstration proposal to appear in IEEE Consumer Communications and Networking Conference, January 8–10, 2006, Las Vegas, http://icampus.mit.edu/projects/publications/Huggable/Huggable-StiehlHuggableCCNC06Finalinfo.pdf. Also Jonathan Mummolo, "Technology: It's a Huggable Robot," *Newsweek*, November 20, 2006, p. 16. Also Cynthia Breazeal, "The Next Best Thing to Being There: Increasing the Emotional Bandwidth of Mediated Communication Using Robotic Avatars," paper presented at MIT Media Lab's H20 conference, Cambridge, MA, May 9, 2007.

7. Breazeal quoted in Clair Bowles, "Expressive Robot Computers," *New Scientist*, March 24, 2006.

8. Cynthia Breazeal and Rosalind Picard, "The Role of Emotion and Inspired Abilities in Relational Robots," in Raja Parasuraman and Matthew Rizzo, *Neuroergonomics: The Brain at Work* (Oxford: Oxford University Press, 2007), pp. 275–91, here, pp. 277–81.

9. Clifford Nass et al., "Improving Automatic Safety by Pairing Driver Emotion and Car Voice Emotion," *Proceedings of the Conference on Human Factors in Computing Systems* (Portland, OR, April 2005): 1973–76.

10. Breazeal and Picard, "Role of Emotion," p. 279.

11. Don Norman, "Cautious Cars and Cantankerous Kitchens: How Machines Take Control," draft introduction to *The Design of Future Things*, unpublished manuscript, http://www.jnd.org/dn.mss/post.html.

12. Kenneth Goodrich, Paul Schutte, and Frank Flemisch, "Application of the H-Mode, A Design and Interaction Concept for Highly Automated Vehicles, to Aircraft," *IEEE* (2006): 6A3-1–6A3-13.

13. Lijin Aryananda, "Art/Sci Collision: Of Human-Robot Bondage," presentation to American Museum of Natural History, April 18, 2007.

14. Interview with Timothy Bickmore, May 2007.

15. Timothy Bickmore et al., "It's Just Like You Talk to a Friend: Relational Agents for Older Adults," *Interacting with Computers* 17, no. 6 (2005): 711–35.

16. Picard quoted in Bennett Daviss, "Tell Laura I Love Her," *New Scientist*, December 3, 2005, pp. 42–46.

17. Sigmund Freud, *The Uncanny*, trans. David McClintock (New York: Penguin Books, 2003), pp. 123–25, 132, 135, 147–48. Originally published as *Das Unheimliche* in 1919 in *Imago* 5, pp. 5–6.

18. E. T. A. Hoffmann, "The Sandman," in *The Tales of Hoffmann*, trans. Michael Bullock (New York: Frederick Ungar, 1963), pp. 1–34.

19. Nass quoted in Daviss, "Tell Laura."

20. Charles Piller, "A Human Touch for Machines," *Los Angeles Times*, March 7, 2002, p. A1.

21. Bowles, "Expressive Robot Computers." Also interviews with Rosalind Picard, August and October 2007.

22. Kathleen Richardson, "Mechanical People," *New Scientist*, June 24, 2006, pp. 56–57.

23. Rosalind Picard and Jonathan Klein, "Computers That Recognize and Respond to User Emotion: Theoretical and Practical Implications," *Interacting with Computers* 14 (2002): 141–69, here, pp. 147–48.

24. Interview with Sherry Turkle, April 2007.

25. Turkle, "Diary."

26. "Producer's Note," in Karel Čapek, *R.U.R. (Rossum's Universal Robots): A Fantastic Melodrama*, trans. Paul Selver (New York: Samuel French, 1923), pp. x–xi.

27. Ivan Klima, *Karel Čapek: Life and Work*, trans. Norma Comrada (North Haven, CT: Catbird Press, 2002), p. 127.

28. David Nye, *Technology Matters: Questions to Live With* (Cambridge, MA: MIT Press, 2006), p. 2.

29. Friedrich Kittler, *Gramophone, Film, Typewriter*, translated and with an introduction by Geoffrey Winthrop-Young and Michael Wutz (Stanford, CA: Stanford University Press, 1999), p. 28.

30. Carolyn Marvin, *When Old Technologies Were New: Thinking about Electric Communication in the Late Nineteenth Century* (New York: Oxford University Press, 1988), pp. 109, 123. Also John Durham Peters, *Speaking into the Air: A History of the Idea of Communication* (Chicago: University of Chicago Press, 1999), p. 91.

31. "The Skinny on IT: The Human Body as a Computer Bus," *Economist*, July 3, 2004, pp. 66–67.

32. Nye, *Technology Matters*, p. 222.

33. Interview with John Halamka, May 2007.

34. Rev. 13:16 (King James Version).

35. John Halamka et al., "A Security Analysis of the VeriChip Implantable RFID Device," *Journal of American Medical Informatics Association* 13 (2006): 601–607.

36. John Halamka, "Straight from the Shoulder," *New England Journal of Medicine* 353, no. 4 (2005): 331–33.

37. Rodney Brooks, *Flesh and Machines: How Robots Will Change Us* (New York: Pantheon Books, 2002), p. 236.

38. *Oxford English Dictionary*, vol. 12 (Oxford: Clarendon Press, 1989), p. 672.

39. Vivian Sobchak, "A Leg to Stand On," in *The Prosthetic Impulse: From a Posthuman Present to a Biocultural Future*, by Maynard Smith and Joanne Morra (Cambridge, MA: MIT Press, 2006), p. 19.

40. N. Katherine Hayles, *How We Became Posthuman* (Chicago: University of Chicago Press, 1999), p. 3. Reprinted in Smith and Morra, *Prosthetic Impulse*, p. 7.

41. Andrew Pollack, "Paralyzed Man Uses Thoughts to Move a Cursor," *New York Times*, July 13, 2006, p. A1. Also Ian Parker, "Reading Minds," *New Yorker*, January 20, 2003, p. 52. Interview with John Donoghue, June 2007.

42. James McGaugh quoted in interview with Jonathan Cott, *On the Sea of Memory: A Journey from Forgetting to Remembering* (New York: Random House, 2005), p. 42.

43. Modafinil user quoted in Graham Lawton, "If I Take a Dose," *New Scientist*, February 18–24, 2006, p. 34.

44. Walter Glannon, "Psychopharmacology and Memory," *Journal of*

Medical Ethics 32 (2006): 74–78. Also Anjam Chatterjee, "The Promise and Predicament of Cosmetic Neurology," *Journal of Medical Ethics* 32 (2006): 110–13. James Vlahos, "Will Drugs Make Us Smarter and Happier," *Popular Science* 267, no. 3 (2005): 64. "Supercharging the Brain," *Economist*, September 18, 2004, pp. 27–29.

45. McGaugh in *On the Sea of Memory*, p. 26.

46. William Shakespeare, *Macbeth*, act 5, scene 3, lines 41–42.

47. Chuck Klosterman, "Amnesia Is the New Bliss," *Esquire* 147, no. 4 (April 2007): 90.

48. David Pearce quoted in Mark White, "Medication Nation," *Ecologist* 35, no. 10 (December 2005–January 2006): 50.

49. *Beyond Therapy: Biotechnology and the Pursuit of Happiness: A Report of the President's Council on Bioethics*, foreword by Leon Kass (Washington, DC: President's Council on Bioethics, 2003), p. 257.

50. N. Katherine Hayles, "The Seductions of Cyberspace," in *Reading Digital Trends*, ed. David Trend (Oxford: Blackwell Publishers, 2001), pp. 305–21, here, p. 313.

51. Quoted in Glannon, "Psychopharmacology and Memory," p. 177.

52. Oliver Sacks, keynote address to MIT Media Lab's H2.0 conference, Cambridge, MA, May 9, 2007.

53. Michael Chorost, *Rebuilt: How Becoming Part Computer Made Me More Human* (Boston: Houghton Mifflin, 2005), p. 16.

54. Ibid., p. 78.

55. Ibid., p. 183.

56. Steve Talbott, *Devices of the Soul: Battling for Our Selves in an Age of Machines* (Sebastopol, CA: O'Reilly Media, 2007), p. vii.

57. Nye, *Technology Matters*, pp. 2–3, 5.

58. Sobchak, "A Leg to Stand On," p. 38.

59. Brooks, *Flesh and Machines*, p. 195.

60. Interview with Steve Talbott, April 2007.

61. Talbott, *Devices of the Soul*, p. ix.

62. Ibid., p. vii.

63. Ibid., pp. 5–9.

64. Ibid., p. 261.

CHAPTER EIGHT

1. Hannah Arendt, *Thinking, Life of the Mind* (New York: Harcourt Brace Jovanovich, 1978), 1:202–10.

2. Paul Saffo, "Six Rules for Effective Forecasting," *Harvard Business Review*, July–August 2007, pp. 122–31.

3. Jane Jacobs, *Dark Age Ahead* (New York: Random House, 2004), pp. 4, 7.

4. Joseph Tainter, *The Collapse of Complex Societies* (Cambridge: Cambridge University Press, 1988), pp. 19–20.

5. Jacobs, *Dark Age Ahead*, p. 4.

6. Strachey quoted in Lewis Lapham, *Waiting for the Barbarians* (London: Verso, 1997), p. 207.

7. Hannah Arendt, *Men in Dark Times* (New York: Harcourt, Brace & World, 1968), p. ix.

8. Daniel Boorstin, *Cleopatra's Nose: Essays on the Unexpected*, ed. Ruth Boorstin (New York: Random House, 1994), p. 7.

9. James McConkey, *The Anatomy of Memory: An Anthology* (New York: Oxford University Press, 1996), p. 5.

10. Ibid., pp. 6–7.

11. Clara Claiborne Park, "The Mother of the Muses: In Praise of Memory," ibid., pp. 173–89.

12. McGaugh quoted in Jonathan Cott, *On the Sea of Memory: A Journey from Forgetting to Remembering* (New York: Random House, 2005), p. 42.

13. Aldous Huxley, *Brave New World and Brave New World Revisited* (New York: Harper Perennial, 2005; first published in 1932 and 1958), pp. 40–41.

14. Eudora Welty, "One Writer's Beginnings," in McConkey, *Anatomy of Memory*, p. 225.

15. Elizabeth Parker, Larry Cahill, and James McGaugh, "A Case of Unusual Autobiographical Remembering," *Neurocase* 12 (2006): 35–49.

16. McGaugh quoted in Farnez Khadem, "UCI Researchers Identify New Form of Superior Memory Syndrome," *University of California/Irvine News Service*, March 13, 2006, http://www.today.uci.edu/news/release _detail.asp?key=1450.

17. James McGaugh, *Memory and Emotion: The Making of Lasting Memories* (New York: Columbia University Press, 2003), p. 131.

18. Jorge Luis Borges, "Funes the Memorious," in *Ficciones* (New York: Grover Press, 1962), pp. 107–15.

19. Eric Kandel, *In Search of Memory: The Emergence of a New Science of Mind* (New York: W. W. Norton, 2006), pp. 59–60, 64–67.

20. Robert Sapolsky, "Stressed Out Memories," *Scientific American Mind* 15, no. 4 (2004): 28–33, here, p. 30.

21. Kandel, *In Search of Memory*, p. 264.

22. Ibid., p. 276.

23. Michael Posner and Mary K. Rothbart, *Educating the Human Brain* (Washington, DC: American Psychological Association, 2007), p. 198.

24. Alberto Manguel, *The Library at Night* (Toronto: A. A. Knopf Canada, 2006), p. 108.

25. James Fallows, "File Not Found," *Atlantic Monthly*, September 2006, p. 142. Also Brad Reagan, "The Digital Ice Age," *Popular Mechanics*, December 2006, p. 95. Matthew Broersma, "It's the End of Your Data as You Know It," *ZDNet*, April 23, 2007, http://resources.zdnet.co.uk/articles/features/0,1000002000,39286796,00.htm.

26. Roy Rosenzweig, "Scarcity or Abundance? Preserving the Past in a Digital Era," *American Historical Review* (June 2003): 735–62, here, p. 741. Jim Barksdale and Francine Berman, "Saving Our Digital Heritage," *Washington Post*, May 16, 2007, p. A15.

27. James Billington, "Is Alexandria Burning Again?" Patricia Doyle Wise lecture, Washington, DC, June 7, 1993. Also Billington quoted in Fallows, "File Not Found."

28. Alexander Stille, *The Future of the Past* (New York: Farrar, Straus and Giroux, 2002), p. 302. Also Daniel Cohen, "The Future of Preserving the Past," *CRM: The Journal of Heritage Stewardship* 12, no. 2 (2005): 6–19.

29. Cohen, "The Future of Preserving the Past," p. 12.

30. Fallows, "File Not Found," p. 142.

31. David Talbot, "The Fading Memory of the State," *Technology Review* 108, no. 7 (2005), p. 44.

32. Clay Shirky, *Library of Congress Archive Ingest and Handling Test Final Report* (Washington, DC: Library of Congress National Digital Information Infrastructure and Preservation Project, 2005). Also interviews with Martha Anderson and Clay Shirky, July 2007.

33. Laura Millar, "Touchstones: Considering the Relationship between Memory and Archives," *Archivaria* 61 (Spring 2006): 105–26, here, p. 114.

34. Jacques Derrida, *Archive Fever: A Freudian Impression* (Chicago: University of Chicago Press, 1996), pp. 1, 91.

35. Ibid., p. 36. Also Mary Bergstein, "Gradiva Medica: Freud's Model Female Analyst as Lizard-Slayer," *American Imago* 60, no. 3 (2003): 285–301.

36. Michele Cloonan, "Whither Preservation?" *Library Quarterly* 71, no. 2 (2001): 231–42, here, p. 235.

37. Eviatar Zerubavel, *Time Maps: Collective Memory and the Social Shape of the Past* (Chicago: University of Chicago Press, 2003), pp. 26, 33.

38. Kandel, *In Search of Memory*, p. 281.

39. McGaugh, *Memory and Emotion*, p. 116.

40. Umberto Eco, "Living in the New Middle Ages," in *Travels in Hyper Reality: Essays*, trans. William Weaver (New York: Harcourt and Brace, 1986), pp. 73–85, here, p. 84.

41. Cloonan, "Whither Preservation?" p. 232.

42. Clive Thompson, "A Head for Detail," *Fast Company*, November 2006, p. 73.

43. Vernor Vinge, "What If the Singularity Does Not Happen?" presentation to Long Now Foundation, San Francisco, CA, February 16, 2007. Also Rosenzweig, "Scarcity or Abundance?" p. 758.

44. Stille, *Future of the Past*, p. 247.

45. Walter Mischel, Yuichi Shoda, and Monica Rodriguez, "Delay of Gratification in Children," *Science* 244, no. 4907 (1989): 933–38.

46. Walter Mischel and Ozlem Ayduk, "Self-Regulation in a Cognitive-Affective Personality System: Attentional Control in the Service of the Self," *Self and Identity* 1 (2002): 113–20, here, p. 115.

47. Arendt, *Thinking*, p. 210.

48. Lisa Tsui, "Fostering Critical Thinking through Effective Pedagogy: Evidence from Four Institutional Case Studies," *Journal of Higher Education* 73, no. 6 (2002): 740.

49. *Measuring Up 2006: The National Report on Higher Education* (Washington, DC: National Center for Public Policy and Higher Education, 2006).

50. George Kuh, "What Student Engagement Data Tells Us about College Readiness," *AAC&U Peer Review* (Winter 2007): 4–8. Also George Kuh, "What We're Learning about Student Engagement from NSSE," *Change* 35, no. 2 (2003): 24.

51. *2007 ACT High School Profile Report: The Graduating Class of 2007* (Iowa City: ACT, 2007), http://www.act.org/news/data/07/index.html.

52. Diana Jean Schemo, "At Two-Year Colleges, Students Eager but Unprepared," *New York Times*, September 2, 2006, p. A1. Also Derek Bok, *Our Underachieving Colleges: A Candid Look at How Much Students Learn and Why They Should Be Learning More* (Princeton, NJ: Princeton University Press, 2006), p. 68.

53. Patricia King quoted in Mark Clayton, "Rethinking Thinking," *Christian Science Monitor*, October 13, 2003, p. 18.

54. Bok, *Our Underachieving Colleges*, pp. 68–69.

55. Ernest Pascarella and Patrick Terenzini, *How College Affects Students: A Third Decade of Research* (San Francisco: John Wiley & Sons, 2005), p. 158.

56. Patricia King and Karen Strohm Kitchener, *Developing Reflective Judgment: Understanding and Promoting Intellectual Growth and Critical Thinking in Adolescents and Adults* (San Francisco: Jossey-Bass Publishers, 1994), pp. 166–67.

57. Barry Kroll, *Teaching Hearts and Minds: College Students Reflect on the Vietnam War in Literature* (Carbondale: Southern Illinois University Press, 1992). Reprinted in King and Kitchener, *Developing Reflective Judgment*, p. 166.

58. Walter Mischel and Ozlem Ayduk, "Willpower in a Cognitive-Affective Processing System," in *Handbook of Self-Regulation*, ed. Roy Baumeister and Kathleen Vohs (New York: Guilford Press, 2004), pp. 99–129.

59. Mischel, Shoda, and Rodriguez, "Delay of Gratification," p. 935.

60. Kathleen Vohs and Roy Baumeister, "Understanding Self-Regulation: An Introduction," in *Handbook of Self-Regulation*, pp. 1–22.

61. Martin Seligman and Angela Duckworth, "Self-Discipline Outdoes IQ in Predicting Academic Performance of Adolescents," *Psychological Science* 16, no. 12 (2005): 939–44.

62. Mischel and Ayduk, "Self-Regulation," p. 118. Also Mischel, Shoda, and Rodriguez, "Delay of Gratification," p. 934.

63. Russell Barkley, "Attention-Deficit/Hyperactivity Disorder and Self-Regulation," in Baumeister and Vohs, *Handbook of Self-Regulation*, pp. 301–23, here, p. 309.

64. Posner and Rothbart, *Educating the Human Brain*, p. 64.

65. Grazyna Kochanska et al., "Individual Differences in Emotionality in Infancy," *Child Development* 69, no. 2 (1998): 375–90.

66. Posner and Rothbart, *Educating the Human Brain*, pp. 86–87.

67. Pascarella and Terenzini, *How College Affects Students*, pp. 186–87.

68. Jennifer Fredericks, Phyllis Blumenfeld, and Alison Paris, "School Engagement: Potential of the Concept, State of the Evidence," *Review of Educational Research* 74, no. 1 (2004): p. 59.

69. American Philosophical Association statement quoted in Peter Facione, *Critical Thinking: A Statement of Expert Consensus for Purposes of Educational Assessment and Instruction* (Millbrae, CA: California Academic Press, 1990). Bok, *Our Underachieving Colleges*, pp. 68–69.

70. Mischel and Ayduk, "Self-Regulation," p. 119.

71. Alexander Astin, *The American Freshman: Thirty-five-Year Trends* (Los Angeles: American Council on Education, Cooperative Institutional Research Program, University of California/Los Angeles, 2002), pp. 16–17.

72. Arthur Levine and Jeanette Cureton, "What We Know about Today's College Students," *About Campus*, March/April 1998, pp. 4–9.

73. Barkley, "Attention-Deficit/Hyperactivity Disorder," pp. 310–16.

CHAPTER NINE

1. William James, *The Principles of Psychology*, ed. Frederick Burkhardt (Cambridge, MA: Harvard University Press, 1981, 1890), 1:381–82.

2. Michael Posner and Marcus Raichle, *Images of Mind* (New York: Scientific American Library, 1994), pp. 182–83.

3. Natalie Boyle et al., "Blindsight in Children: Does It Exist and Can It Be Used to Help the Child?" *Development of Medicine and Child Neurology* 47, no. 10 (2005): 699.

4. Michael Posner, "Genes and Experience Shape Brain Networks of Conscious Control," *Progress in Brain Research* 150 (2005): 173–83, here, p. 180. Also Michael Posner and Mary K. Rothbart, *Educating the Human Brain* (Washington, DC: American Psychological Association, 2007), pp. 60–61.

5. Posner and Rothbart, *Educating the Human Brain*, pp. 68–69.

6. Naomi Eilen, ed., *Joint Attention: Communication and Other Minds: Issues in Philosophy and Psychology* (Oxford: Clarendon Press, 2005), p. 1.

7. Marie Rocha, Laura Schreibman, and Aubyn Stahmer, "Effectiveness of Training Parents to Teach Joint Attention in Children with Autism," *Journal of Early Intervention* 29, no. 2 (2007): 154.

8. Interview with Bruce McCandliss, October 2007.

9. Michael Posner and Stephen Boies, "Components of Attention," *Psychological Review* 78 (1971): 391–408.

10. Amir Raz and Jason Buhle, "Typologies of Attention Networks," *Nature Reviews Neuroscience* 7 (May 2006): 367–79.

11. Posner and Rothbart, *Educating the Human Brain*, p. 93.

12. Michael Posner and Mary K. Rothbart, "Research on Attention Networks as a Model for the Integration of Psychological Science," *Annual Review of Psychology* 58 (2007): 1–23.

13. Michael Posner, "How I Got Here," in *Developing Individuality in the Human Brain: A Tribute to Michael I. Posner*, ed. Ulrich Mayr, Edward Awh, and Steven Keele (Washington, DC: American Psychological Association, 2005), pp. 237–46.

14. Andrea Berger, Gabriel Tzur, and Michael Posner, "Infant Brains Detect Arithmetic Errors," *Proceedings of the National Academy of Sciences* 103, no. 33 (2006): 12649–53.

15. M. Rosario Rueda et al., "Training, Maturity and Genetic Influences on the Development of Executive Attention," *Proceedings of the National Academy of Sciences* 102, no. 41 (2005): 14931–36. Also Raz and Buhle, "Typologies of Attention Networks."

16. Michael Posner, "Developing Brain Mechanisms of Self-Regulation" (presentation to Mount Sinai School of Medicine Psychiatry Department, March 20, 2007).

17. Steven Haggbloom et al., "Eminent Psychologists of the Twentieth Century," *Review of General Psychology* 6, no. 2 (2002): 139–52.

18. Nicholas Schiff and Michael Posner, "Consciousness: Mechanisms, Keynote," in *Encyclopedia of Life Sciences* (New York: John Wiley & Sons, 2005), pp. 1–6. Also Michael Posner, "Measuring Alertness," in *Molecular and Biophysical Mechanisms of Arousal, Alertness and Attention*, ed. Don Pfaff and Brigitte Kieffer (New York: New York Academy of Sciences, in press).

19. Schiff and Posner, "Consciousness: Mechanisms," p. 3.

20. Jeffrey Halperin and Kurt Schulz, "Revisiting the Role of the Prefrontal Cortex in the Pathophysiology of Attention-Deficit/Hyperactivity Disorder," *Psychological Bulletin* 132, no. 4 (2006): 560–81.

21. Schiff and Posner, "Consciousness: Mechanisms," p. 5.

22. Raz and Buhle, "Typologies of Attention Networks," p. 374.

23. Ignas Skrupskelis, "William James," *American National Biography Online*, February 2000, http://www.anb.org/articles/20/20-01725. Also Linda Simon, *Genuine Reality: A Life of William James* (New York: Harcourt Brace, 1998).

24. Linda Simon, ed., *William James Remembered* (Lincoln: University of Nebraska Press, 1996), p. 119.

25. Ibid., p. 37.

26. William James, *Talks to Teachers on Psychology and to Students on Some of Life's Ideals* (New York: Henry Holt and Company, 1906), p. 113.

27. Interview with Clifford Saron, May 2007.

28. James, *Talks to Teachers*, p. 65.

29. Joan McDowd et al., "Attentional Abilities and Functional Outcomes Following Stroke," *Journals of Gerontology: Series B: Psychological Sciences and Social Sciences* 58B, no. 1 (2003): 45–53.

30. Sharon Begley, *Train Your Mind, Change Your Brain: How a New Science Reveals Our Extraordinary Potential to Transform Ourselves* (New York: Ballantine Books, 2007), pp. 6, 8.

31. Ibid., pp. 74, 91, 100.

32. Stanley Colcombe et al., "Cardiovascular Fitness, Cortical Plasticity, and Aging," *Proceedings of the National Academy of Sciences* 101, no. 9 (2004): 3316–21.

33. Helen Neville quoted in Begley, *Train Your Mind*, pp. 90–92.

34. Bstan-dzin-rgya-mtsho, the Dalai Lama XIV, *The Universe in a Single Atom: The Convergence of Science and Spirituality* (New York: Morgan Road Books, 2005), p. 150.

35. Huston Smith, *The Religions of Man* (New York: Harper & Row, 1958, 1964), pp. 96–97.

36. B. Alan Wallace, *Contemplative Science: Where Buddhism and Neuroscience Converge* (New York: Columbia University Press, 2007), p. 136.

37. Interviews with Alan Wallace, May and July 2007.

38. B. Alan Wallace, *The Attention Revolution: Unlocking the Power of the Focused Mind* (Boston: Wisdom Publications, 2006), p. 15.

39. Interview with Cass McLaughlin, May 2007.

40. Amishi Jha, Jason Krompinger, and Michael Baime, "Mindfulness Training Modifies Subsystems of Attention," *Cognitive, Affective and Behavioral Neuroscience* 7, no. 2 (2007): 109–19.

41. David Washburn, "Picking up the Check When It's Time to Pay

Attention," *Psychological Science Agenda* 20, no. 5 (2006): 1–4. Duane Rumbaugh and David Washburn, "Attention and Memory in Relation to Learning," in *Attention, Memory and Executive Function*, ed. G. Reid Lyon and Norman Krasnegor (Baltimore, MD: Paul H. Brookes, 1995), pp. 199–219. Also interview with David Washburn, September 2007.

42. M. Rosario Rueda, Mary K. Rothbart, and Michael Posner, "Training, Maturation and Genetic Influences on the Development of Executive Attention," *Proceedings of the National Academy of Sciences* 102, no. 41 (2005): 14931–36.

43. Posner quoted in Bridget Murray, "Training Young Minds Not to Wander," *Monitor on Psychology* 34, no. 9 (2003): 58.

44. Leanne Tamm et al., "Can Attention Itself Be Trained: Attention Training for Children At-Risk for ADHD," in *Attention Deficit/Hyperactivity Disorder: A 21st Century Perspective*, ed. Keith McBurnett et al. (New York: Marcel Dekler, 2007).

45. Torkel Klingberg, "Computerized Training of Working Memory in Children with ADHD—A Randomized, Controlled Trial," *Journal of American Academy of Child and Adolescent Psychiatry* 44, no. 2 (2005): 177–86. Also interview with Dr. Christopher Lucas, associate professor of child and adolescent psychiatry at New York University, August 2007.

46. Interview with Leanne Tamm, August 2007. Also Tamm et al., "Can Attention Itself Be Trained."

47. "Greater love hath no man than this, that a man lay down his life for his friends," John 15:13 (King James Version).

48. *Oxford English Dictionary*, vol. 4 (Oxford: Clarendon Press, 1989), p. 863.

49. Melanie Thernstrom, "My Pain, My Brain," *New York Times Magazine*, May 14, 2006, pp. 51–55.

50. Hunter Hoffman, "Virtual Reality Therapy," *Scientific American*, August 2004, pp. 58–65.

51. James, *Principles of Psychology*, 2:992.

52. William Shakespeare, *Anthony and Cleopatra*, act III, scene vii, 77.

53. John Henry Blunt, ed., *The Myroure of Oure Ladye*, Early English Text Society ed. (London: N. Trubner & Co., 1873), pp. 41–52.

54. Jonathan Smallwood and Jonathan Schooler, *Psychological Bulletin* 132, no. 6 (2006): 946–58. Also Jonathan Smallwood, Daniel Fishman, and Jonathan Schooler, "Counting the Cost of an Absent Mind: Mind Wandering

as an Underrecognized Influence on Educational Performance," *Psychonomic Bulletin & Review* 14, no. 2 (2007): 230.

55. Interviews with Jacob Collins, July and September 2007.

56. James Panero, "The New Old School," *New Criterion*, September 2006, pp. 104–107.

57. Jin Fan et al., "Testing the Efficiency and Independence of Attentional Networks," *Journal of Cognitive Neuroscience* 14, no. 34 (2002): 340–47.

INDEX